SOMATIC CELL GENETICS OF WOODY PLANTS

FORESTRY SCIENCES

Baas P, ed: New Perspectives in Wood Anatomy. 1982. ISBN 90-247-2526-7
Prins CFL, ed: Production, Marketing and Use of Finger-Jointed Sawnwood. 1982. ISBN 90-247-2569-0
Oldeman RAA, et al., eds: Tropical Hardwood Utilization: Practice and Prospects. 1982. ISBN 90-247-2581-X
Den Ouden P and Boom BK: Manual of Cultivated Conifers: Hardy in Cold and Warm-Temperate Zone. 1982. ISBN 90-247-2148-2
Bonga JM and Durzan DJ, eds: Tissue Culture in Forestry. 1982. ISBN 90-247-2660-3
Satoo T and Magwick HAI: Forest Biomass. 1982. ISBN 90-247-2710-3
Van Nao T, ed: Forest Fire Prevention and Control. 1982. ISBN 90-247-3050-3
Douglas J: A Re-appraisal of Forestry Development in Developing Countries. 1983. ISBN 90-247-2830-4
Gordon JC and Wheeler CT, eds: Biological Nitrogen Fixation in Forest Ecosystems: Foundations and Applications. 1983. ISBN 90-247-2849-5
Németh MV: The Virus-Mycoplasma and Rikettsia Disease of Fruit Trees. ISBN 90-247-2868-1
Duryea ML and Landis TD, eds: Forest Nursery Manual: Production of Bareroot Seedlings. 1984. ISBN 90-247-2913-0
Hummel FC, ed: Forest Policy: A Contribution to Resource Development. 1984. ISBN 90-247-2883-5
Manion PD, ed: Scleroderris Canker of Conifers. 1984. ISBN 90-247-2912-2
Duryea ML and Brown GN, eds: Seedling Physiology and Reforestation Success. 1984. ISBN 90-247-2949-1
Staaf KAG and Wiksten NA: Tree Harvesting Techniques. 1984. ISBN 90-247-2994-7
Boyd JD: Biophysical Control of Microfibril Orientation in Plant Cell Walls. 1985. ISBN 90-247-3101-1
Findlay WPK, ed: Preservation of Timber in the Tropics. 1985. ISBN 90-247-3112-7
Samset I: Winch and Cable Systems. 1985. ISBN 90-247-3205-0
Leary RA: Interaction Theory in Forest Ecology and Management. 1985. ISBN 90-247-3220-4
Gessel SP: Forest Site and Productivity. 1986. ISBN 90-247-3284-0
Hennessey TC, Dougherty PM, Kossuth SV and Johnson JD, eds: Stress Physiology and Forest Productivity. 1986. ISBN 90-247-3359-6
Shepherd KR: Plantation Silviculture. 1986. ISBN 90-247-3379-0
Sohlberg S and Sokolov VE, eds: Practical Application of Remote Sensing in Forestry. 1986. ISBN 90-247-3392-8
Bonga JM and Durzan DJ, eds: Cell and Tissue Culture in Forestry. Volume 1: General Principles and Biotechnology. 1987. ISBN 90-247-3430-4
Bonga JM and Durzan DJ, eds: Cell and Tissue Culture in Forestry. Volume 2: Specific Principles and Methods: Growth and Developments. 1987. ISBN 90-247-3431-2
Bonga JM and Durzan DJ, eds: Cell and Tissue Culture in Forestry. Volume 3: Case Histories: Gymnosperms, Angiosperms and Palms. 1987. ISBN 90-247-3432-0
(Set Vols. 1–3: ISBN 90-247-3433-9)
Richards EG, ed: Forestry and the Forest Industries: Past and Future. 1987. ISBN 90-247-3592-0
Kossuth SV and Ross SD, eds: Hormonal Control of Tree Growth. 1987. ISBN 90-247-3621-8
Sundberg U and Silversides CR, eds: Operational Efficiency in Forestry. Volume 1: Analysis. 1988. ISBN 90-247-3683-8

Somatic Cell Genetics of Woody Plants

Proceedings of the IUFRO Working Party S2. 04-07 Somatic Cell Genetics, held in Grosshansdorf, Federal Republic of Germany, August 10–13, 1987

edited by

M.R. AHUJA

Federal Research Centre for Forestry and Forest Products, Institute of Forest Genetics and Forest Tree Breeding, Grosshansdorf, F.R.G.

KLUWER ACADEMIC PUBLISHERS

DORDRECHT / BOSTON / LONDON

Library of Congress Cataloging in Publication Data

Somatic cell genetics of woody plants : proceedings of the IUFRO Working Party S2.04-07 "Somatic Cell Genetics", held in Grosshansdorf, Federal Republic of Germany, 10-13 August 1987 / edited by M.R. Ahuja.
p. cm. -- (Forestry sciences)
ISBN 9024737281
1. Trees--Propagation--In vitro--Congresses. 2. Trees--Cytogenetics--Congresses. 3. Trees--Biotechnology--Congresses. 4. Woody plants--Propagation--In vitro--Congresses. 5. Woody plants--Cytogenetics--Congresses. 6. Woody plants--Biotechnology--Congresses. I. Ahuja, M. R., 1933- . II. IUFRO Working Party S2.04-07 "Somatic Cell Genetics". III. Series.
SD403.5.S65 1988
634.9'56--dc19 88-15078
CIP

ISBN 90-247-3728-1

Published by Kluwer Academic Publishers,
P.O. Box 17, 3300 AA Dordrecht, The Netherlands.

Kluwer Academic Publishers incorporates
the publishing programmes of
D. Reidel, Martinus Nijhoff, Dr W. Junk and MTP Press.

Sold and distributed in the U.S.A. and Canada
by Kluwer Academic Publishers,
101 Philip Drive, Norwell, MA 02061, U.S.A.

In all other countries, sold and distributed
by Kluwer Academic Publishers Group,
P.O. Box 322, 3300 AH Dordrecht, The Netherlands.

Printed in The Netherlands

TABLE OF CONTENTS

Participants (Photo) xiv

WELCOME ADDRESS - D. Noack xvii

OPENING REMARKS - G.H. Melchior xix

INTRODUCTION - M.R. Ahuja xxi

SOMATIC EMBRYOGENESIS

DEVELOPMENT AND CHARACTERIZATION OF *IN VITRO* EMBRYOGENIC SYSTEMS IN CONIFERS - M.R. Becwar, S.R. Wann, M.A. Johnson, S.A. Verhagen, R.P. Feirer and R. Nagmani 1

1. ABSTRACT 1
2. INTRODUCTION 1
3. INITIATION OF EMBRYOGENIC CALLUS IN CONIFERS 2
3.1. *Picea* 2
3.1.1. Origin of embryogenic callus 2
3.1.2. Optimum initiation window 2
3.1.3. Quantification of embryogenic capacity 4
3.2. *Pinus* 5
3.2.1. Pond pine 5
3.2.2. Loblolly pine 7
3.2.3. White pine 7
3.3. Comparison of *in vitro* embryogenesis in *Pinus* and *Picea* 7
4. BIOCHEMICAL CHARACTERIZATION OF EMBRYOGENIC CONIFER CALLUS 9
5. ULTRASTRUCTURAL CHARACTERIZATION OF EMBRYOGENIC CONIFER CALLUS 10
6. CONIFER EMBRYOGENIC SUSPENSION CULTURE 12
7. DEVELOPMENT OF CONIFER SOMATIC EMBRYOS TO PLANTS 12
8. COMPARISON OF *IN VIVO* AND *IN VITRO* CONIFER EMBRYOGENESIS 14
9. SUMMARY 14

STUDIES ON EMBRYOGENESIS OF WOODY PLANTS IN CHINA - Chen Zhenghua, Yao Yuguang and Zhang Lihua 19

0. ABSTRACT 19
1. INTRODUCTION 19
2. FACTORS AFFECTING EMBRYOID FORMATION 19
2.1. Pretreatment 19
2.2. Explant 20
2.3. Media and its supplements 20
2.4. Environmental factors 20
3. ORIGIN OF EMBRYOIDS 21
3.1. Embryoids from pollen 21
3.2. Embryoids from somatic tissue 21
4. ABNORMAL EMBRYOIDS AND MEASURES TO PREVENT THEIR FORMATION 22
4.1. Abnormal embryoid types 22
4.1.1. Abnormal globular, heart, torpedo, and rod shaped 22
4.1.2. Embryoids from multiple meristems 22

4.1.3. Embryoids with abnormal root and shoot poles 22
4.1.4. Embryoids with abnormal cotyledons 22
4.1.5. Vitreous embryoids 22
4.2. Measures to prevent formation of abnormal embryoids 22
4.2.1. Embryoid maturation 22
4.2.2. Hormone adjustments 23
4.2.3. Removal of toxins 23
4.2.4. Removal of abnormal embryoids from culture 23

MORPHOLOGICAL DEFINITION OF PHENOCRITICAL PERIOD FOR INITIATION OF HAPLOID EMBRYOGENIC TISSUE FROM EXPLANTS OF LARIX DECIDUA - P. von Aderkas and J.M. Bonga 29

ABSTRACT 29
1. INTRODUCTION 29
2. PROCEDURE 30
2.1. Materials and methods 30
2.1.1. Description of source material 31
2.1.2. Establishment of morphological markers 31
2.1.3. Application of morphological markers 32
2.1.4. Degree-days 32
2.1.5. Megagametophyte position 32
2.1.6. Statistics 32
2.1.7. Induction 32
3. RESULTS 32
3.1.1. Morphological markers of Larix decidua 32
3.1.2. Predictive value of morphological markers for Larix 33
3.1.3. Predictive value of morphological markers of Larix decidua on inducibility of Picea glauca 34
3.1.4. Megagametophyte position in Larix decidua 34
4. DISCUSSION 36

PRODUCTION OF HAPLOID PLANTLETS IN CULTURES OF UNPOLINATED OVULES OF HEVEA BRASILIENSIS MUELL.-ARG - Chen Zhenghua, Li Wenbin, Zhang Lihua, Xu Xuen and Zhang Shijie 39

0. ABSTRACT 39
1. INTRODUCTION 39
2. MATERIALS AND METHODS 39
2.1. Explant 39
2.2. Surface sterlization 40
2.3. Excision and inoculation 40
2.4. Culture media 40
2.4.1. Dedifferentiation medium 40
2.4.2. Differentiation medium 40
2.4.3. Plant forming medium 40
2.4.4. Environmental factors 40
2.5. Cytological procedures 41
3. RESULTS 42
4. DISCUSSION

SOMATIC EMBRYOGENESIS IN TISSUE CULTURES OF WALNUT (*JUGLANS NIGRA*, *J. MAJOR* AND HYBRIDS *J. NIGRA* X *J. REGIA*) - D. Cornu 45

ABSTRACT 45
1. INTRODUCTION 45
2. MATERIAL AND METHODS 45
3. RESULTS AND DISCUSSION 46

IN VITRO EMBRYOGENIC CALLUS FORMATION IN *CHIMONANTHUS* - L. Radojevic, N. Djordjevic and M. Guc-Scekic 51

PLANT REGENERATION OF HORSE CHESTNUT BY IN VITRO CULTURE - L. Radojevic 53

GENETIC TRANSFORMATION

A MODEL SYSTEM FOR GENE TRANSFER IN CONIFERS: EUROPEAN LARCH AND *AGROBACTERIUM* - D.F. Karnosky, A.M. Diner and W.M. Barnes 55

1. ABSTRACT 55
2. INTRODUCTION 56
3. MICROPROPAGATION 56
4. GENE TRANSFER 57
5. POTENTIAL FOR GENETIC IMPROVEMENT 59

REGENERATION AND TRANSFORMATION OF APPLE PLANTS USING WILD-TYPE AND ENGINEERED PLASMIDS IN AGROBACTERIUM SPP. - D.J. James, A.J. Passey, S. Predieri and E. Rugini 65

ABSTRACT 65
1. INTRODUCTION 65
2. PROCEDURE 66
2.1. Materials and methods 66
2.1.1. Micropropagation and regeneration 66
2.1.2. Transformation procedures 66
2.2. Statistical treatments 67
3. RESULTS AND DISCUSSION 67
3.1. Regeneration from complex explants - somaclonal variation 67
3.2. Transformation with *Agrobacterium* spp. 68
3.2.1. *Agrobacterium tumefaciens* - wild type plasmids 68
3.2.2. *Agrobacterium tumefaciens* - disarmed engineered plasmids - binary vectors 68
3.2.3. *Agrobacterium rhizogenes* co-culture 69
3.2.4. Shoot inoculation 69

EXPRESSION OF AN HERBICIDE TOLERANCE GENE IN YOUNG PLANTS OF A TRANSGENIC HYBRID POPLAR CLONE - D.E. Riemenschneider, B.E. Haissig, J. Sellmer and J.J. Fillatti 73

ABSTRACT 73
1. INTRODUCTION 73
2. MATERIALS AND METHODS 74

2.1. Genetic transformation 74
2.2. *In vitro* propagation 75
2.3. Roundup spray tests 75
2.4. *Agrbacterium* assay 76
3. RESULTS 76
3.1. Roundup spray tests 76
3.2. *Agrobacterium* assay 78
4. DISCUSSION 78

TRANSFORMATION OF HYBRID *POPULUS TREMULA* X *P. ALBA* BY *AGROBACTERIUM TUMEFACIENS* - M.F. Michel, F. Delmotte and C. Depierreux 81

GENE TRANSFER IN WOODY PLANTS: PERSPECTIVES AND LIMITATIONS - M.R. Ahuja 83

0. ABSTRACT 83
1. INTRODUCTION 83
2. HYBRIDIZATION 84
2.1. Backcross 84
3. FUSION OF PROTOPLASTS 86
4. *AGROBACTERIUM* PLASMID-VECTOR SYSTEM 87
4.1. Biology 87
4.1.1. The Ti plasmid 87
4.1.2. The genes in T-DNA 87
4.1.3. The virulence region 87
4.1.4. The T-DNA borders 88
4.2. Vectors for gene transfer 88
4.2.1. Selectable marker genes 89
5. TRANSFORMATION IN TREE SPECIES 90
5.1. Tumor formation and growth autonomy 90
5.2. Gene transfer with selectable markers 90
6. TRANSFER OF FOREIGN GENES WITHOUT *AGROBACTERIUM* MEDIATION 90
6.1. The delivery system 90
6.2. Direct gene transfer into protoplasts 91
6.3. Direct injection of DNA 91
6.3.1. Microinjection 91
6.3.2. Injection into plants 91
7. GENETICS OF TRANSGENIC PLANTS 92
7.1. Inheritance of selectable markers 92
7.2. Somatoclonal variation 94
8. PERSPECTIVES AND LIMITATIONS 94

GENETIC CONTROL OF MORPHOGENESIS

SOMATIC CELL GENETIC RESEARCH IN FORESTRY: INTEGRATION OF CYTOGENETICS, TISSUE CULTURE, AND MOLECULAR GENETICS - S.E. Schlarbaum 103

ABSTRACT 103
1. INTRODUCTION 103
2. REVIEW 104
2.1. Somatic cell hybridization 104
2.2. Chromosome- and microcell-mediated gene transfer 105
2.3. Flow cytogenetics 107

3. DISCUSSION 108
3.1. Gene mapping 108
3.1.1. Biochemical markers and genetic probes 108
3.1.2. Restriction fragment length polymorphisms (RFLP) 108
3.1.3. Saturated linkage maps 110
3.2. Somatic cell genetics and tree improvement 111
4. CONCLUDING REMARKS 112

DIFFERENTIAL NORMS OF REACTION IN TISSUE CULTURE OF BIRCH - Heide Glock 119

ABSTRACT 119
1. INTRODUCTION 119
2. MATERIAL 119
3. METHODS 120
4. CHARACTERS 120
5. MODEL 121
6. RESULTS 122
7. DISCUSSION 124

DETERMINATION OF PLANTLET REGENERATION CAPACITY OF SELECTED ASPEN CLONES *IN VITRO* - M.R. Ahuja, D. Krusche and G.H. Melchior 127

0. ABSTRACT 127
1. INTRODUCTION 127
2. MATERIAL AND METHODS 128
2.1. Culture of bud meristems 128
2.2. Statistical analysis 128
3. RESULTS 129
3.1. Response of MS to BA-treatments 129
3.2. Response of MS to families and BA-treatments 130
3.3. Correlation of MS induction with other morphological and physiological traits 130
4. DISCUSSION 131
4.1. Optimal BA concentration for MS induction 131
4.2. Differential response of families to BA treatments 132
4.3. Correlation between MS and other traits 132
4.4. Aging and micropropagation 133
4.5. Approval and release of clones into practice 133

SUSPENSION CULTURE OF DIPTEROCARP *SHOREA ROXBURGHII* G. DON. - D.S Gunasekara, E.S. Scott and A.N. Rao 137

ABSTRACT 137
1. INTRODUCTION 137
2. MATERIALS AND METHODS 137
2.1. Cell culture 137
2.2. Stock culture 138
2.3. Growth curves 138
2.4. Plating on solid medium 138
2.5. Observations 138
3. RESULTS AND DISCUSSION 138
3.1 Establishment and growth of suspension cultures 138

3.2. Plating suspension on agar medium 140
3.3. Observation on cells in suspension 140
4. CONCLUSIONS 140

ISOZYME ANALYSIS OF MICROPROPAGATED BETULA PENDULA ROTH. - Grzegorz Kosinski, Amanda M. Gordon and Ian R. Brown 143

1. INTRODUCTION 143
2. PROCEDURE 143
2.1. Materials and methods 143
3. RESULTS 143
4. DISCUSSION 144

TISSUE CULTURE STUDIES OF MAHOGANY TREE, SWEITENIA - S. Venketeswaran , M.A.D.L. Dias, S. Sultanbawa and U.V. Weyers 147

1. ABSTRACT 147
2. INTRODUCTION 147
3. MATERIALS AND METHODS 148
3.1. Seed material 148
3.2. Methods 148
3.2.1. Callus initiation 148
3.2.2. Culture media 148
3.2.3. Micropropagation 148
4. RESULTS 148
4.1. General growth response 148
4.2. Response of cotyledons 148
4.3. Response of stem segments and leaf tissue 149
4.4. Progress in micropropagation 149
5. CONCLUSIONS AND PROSPECTS 150

ATTEMPTS TO MICROPROPAGATE MATURE LARIX DECIDUA MILL. - J.M. Bonga and P. von Aderkas 155

ABSTRACT 155
1. INTRODUCTION 155
2. MATERIALS AND METHODS 156
2.1. Collection and storage 156
2.1.1. Female cones at around the time of meiosis 156
2.1.2. Vegetative and sexual buds 156
2.2. Surface sterlization and excision 157
2.2.1. Female cones at about the time of meiosis 157
2.2.2. Vegetative and sexual buds 157
2.3. Media and culture conditions 157
3. RESULTS 157
3.1. Description of shoot and root development 157
3.1.1. Female cones at about the time of meiosis 157
3.1.2. Vegetative and sexual buds 159
3.2. The effect of some experimental variables 160
3.2.1. Recording of data 160
3.2.2. Experiments with female cones collected when at or near meiosis 160
3.2.3. Experiments with vegetative and sexual buds 161
4. DISCUSSION 161

PROTOPLAST CULTURE OF WOODY ROSACEAE AND A COMPARISON TO HERBACEOUS ROSACEAE - Jörg Jörgensen and Horst Binding 169

0. ABSTRACT 169
1. INTRODUCTION 169
2. MATERIALS AND METHODS 169
3. RESULTS AND DISCUSSION 169

DIFFERENTIAL GROWTH RESPONSE OF ASPEN CLONES STORED AT SUB-ZERO TEMPERATURES - M.R. Ahuja 173

0. ABSTRACT 173
1. INTRODUCTION 173
2. MATERIALS AND METHODS 174
2.1. Species and storage temperatures 174
2.1.1. Storage of twigs 174
2.1.2. Storage of buds 174
2.1.3. *In vitro* culture 174
3. RESULTS AND DISCUSSION 174
3.1. Storage of twigs 174
3.1.1. Storage for one month 176
3.1.2. Storage for one year 176
3.1.3. Storage for two years 177
3.2. Storage of dormant buds 179
3.2.1. Storage for five weeks 179
3.2.2. Storage for one year 179
3.2.3. Storage for two years 179

IN VITRO SCREENING, SOMACLONAL SELECTION, TESTING AND DEPLOYMENT

SCREENING FOR RESISTANCE TO *HYPOXYLON MAMMATUM* IN *POPULUS TREMULOIDES* CALLUS AND MICROPROPAGATED PLANTLETS - F. Valentine, S. Baker, R. Belanger, P. Manion and D. Griffin 181

INCREASED HERBICIDE TOLERANCE OF *IN VITRO* SELECTED HYBRID POPLAR - C.H. Michler and B.E. Haissig 183

ABSTRACT 183
1. INTRODUCTION 183
2. MATERIALS AND METHODS 184
2.1. Plant material 184
2.2. Establishment of continuous shoot cultures 184
2.3. Optimization of adventitious shoot production 185
2.4. Somaclonal variant selection 185
2.5. Determination of herbicide lethality 185
2.6. Rechallenge of putative herbicice tolerant somaclones 185
2.7. Preliminary ex vitro testing of somaclonal variants 185
3. RESULTS 186
4. DISCUSSION 187

THE OCCURRENCE OF ENDOGENOUS BACTERIA IN BIRCH SHOOT CULTURES - Amanda M. Gordon and Ian R. Brown 191

1. INTRODUCTION 191
2. PROCEDURE 191
2.1. Materials and methods 191
3. RESULTS 191
4. DISCUSSION 193

ESTABLISHMENT OF A STERILE FOAM CULTIVATION SYSTEM: ELIMINATION OF FORMALDEHYDE - K. Gebhardt 195

ABSTRACT 195
1. INTRODUCTION 195
2. PROCEDURE 196
2.1. Equipment and materials 196
2.2. Preparation of liquid nutrient solution 196
2.3. Preparation of the foam cultivation system and foam pretreatment 197
2.4. Evaluation of effective formaldehyde elimination 197
2.5. Evaluation of pH shift 197
3. RESULTS 197
4. APPLICATION 198
4.1. Cultivation of shoot tip cultures with pretreated foam 198

TESTING AND DEPLOYMENT OF BRAVE NEW PLANTLINGS - W.J. Libby 201

0. ABSTRACT 201
1. INTRODUCTION 201
2. PROBLEMS WITH DEPLOYMENT 201
2.1. Forestry is a conservative profession, and rightly so 201
2.2. Conservative foresters demand genetic diversity 201
3. PROBLEMS WITH TESTING 202
3.1. Time 202
3.2. Total performance 202
3.3. Number and characteristics of test sites 202
3.4. Propagule comparability 203
4. PLANT MATERIAL USED DURING FUNDAMENTAL RESEARCH AND TECHNIQUE DEPLOYMENT 203
5. RESPONSE TO A QUESTIONAIRE 203
5.1. Testing and deployment 203
5.2. Genetic identification 206
5.3. Genetic diversity 206
5.4. Whole-tree collaborators 206
5.5. Respondents' organizations 207
6. A FEW SUGGESTIONS 207
6.1. Provenance 207
6.2. Pedigree 207
6.3. Clones 207
6.4. The ideal 207
6.5. Minority mixes 208
7. SOME WORRIES ABOUT PROFESSIONAL INFRASTRUCTURE 208

RULES GOVERNING THE RELEASE OF FOREST REPRODUCTIVE MATERIAL DERIVED BY *IN VITRO*-CULTURE - H.J. Muhs 211

0. ABSTRACT 211
1. INTRODUCTION 211
2. THE PRINCIPLE OF APPROVAL 212
3. THE PROBLEM OF APPROVING *IN VITRO*-CULTURE DERIVED PLANT MATERIAL AND SOLUTION PROPOSED 214
4. RULES FOR THE MARKETING OF MULTICLONAL MIXTURES 216
5. OUTLOOK 219

GUIDELINES FOR RESEARCH AND DEVELOPMENT IN BIOTECHNOLOGY - Stanley L. Krugman 221

ABSTRACT 221
1. INTRODUCTION 221
2. PHILOSOPHY OF THE RESEARCH GUIDELINES 223
3. REVIEW PROCESS 223
4. CONTAINMENT 224
5. RELEASE 224
6. CONCLUSION 225

IUFRO

IUFRO WORKSHOP

SOMATIC CELL GENETICS OF WOODY PLANTS

10-13 August, 1987

Grosshansdorf, FRG

1. Fredrick Valentine (USA)
2. William J. Libby (USA)
3. Karl Gebhardt (FRG)
4. Patrick von Aderkas (Canada)
5. Roger Timmis (USA)
6. Jan M. Bonga (Canada)
7. Dietmar Krusche (FRG)
8. Thomas Geburek (FRG)
9. Otto Tzschacksch (GDR)
10. Walter Kunkel (FRG)
11. M. Raj Ahuja (FRG)
12. Heide Glock (FRG)
13. Daniel Cornu (France)
14. Howard B. Kriebel (USA)
15. Armin König (FRG)
16. David Karnosky (USA)
17. Hans J. Muhs (FRG)
18. Helmut Schenck (FRG)
19. Jörg Jörgensen (FRG)
20. Erich Dold (FRG)
21. E. Tsang (Canada)
22. Karin Meier (FRG)
23. B. R. Stephan (FRG)
24. Marie-France Michel (France)
25. David James (UK)
26. G. Kosinski (Poland)
27. G. Heinrich Melchior (FRG)
28. Chen Zhenghua (China)
29. Jerry Pullman (USA)
30. Charles Michler (USA)
31. Y. Han (China)
32. Gevenna Djaraevic (Yugoslavia)
33. Amanda Gordon (UK)
34. Scott Schlarbaum (USA)
35. Duhem (Belgium).

WELCOME ADDRESS

Ladies and gentlemen, distinguished guests, and dear colleagues.

On behalf of the Federal Research Centre for Forestry and Forest Products, it is a great honour and pleasure for me to welcome you very heartily at our Institute of Forest Genetics and Forest Tree Breeding. We are glad that you have chosen our Institute in Grosshansdorf for the first meeting of your IUFRO Working Party of the Subject Group S2.04-07 'Somatic Cell Genetics'. This Working Party was created at the last IUFRO World Congress held in 1986 in Ljubjana, Yugoslavia, and Dr. Ahuja, the organizer of this Workshop, was elected Chairman of this new Working Party.

I take as a sign of importance that a Workshop on the Somatic Cell Genetics of Woody Plants has been arranged so early after the last IUFRO World Congress. There is a lot of interest in biotechnology and its application to tree species. Some of the main areas of biotechnology, for example, somatic embryogenesis, micropropagation, cell manipulation, genetic transformation, in vitro screening, germplasm preservation, testing and deployment, and guidelines for research in biotechnology are addressed at this International Workshop by well known scientists like you coming from different parts of the world. I am sure you will have an interesting and a successful meeting.

I hope that at least some of you will have an opportunity to visit other Institutes of our Federal Research Centre to become familiar with the subject matter of research going on there. Briefly, I may mention that the Federal Research Centre for Forestry and Forest Products consists of 6 Institutes:

- Institute of World Forestry and Ecology
- Institute of Forest Genetics and Forest Tree Breeding
- Institute of Economy
- Institute of Wood Biology and Wood Preservation
- Institute of Wood Chemistry and Chemical Wood Technology, and
- Institute of Wood Physics and Mechanical Wood Technology.

All the other five Institutes of our research organization are located about 25 km from here, at Hamburg-Lohbrügge.

Our Federal Research Centre belongs to the Federal Ministry of Food, Agriculture and Forestry, Bonn. In this centre predominantly applied but also basic research is carried out in the fields of forestry and wood sciences.

Dr. Melchior, the Director of the Instiute of Forest Genetics and Forest Tree Breeding, will briefly inform you regarding the research programs of this Institute.

Once again I welcome you at the Federal Research Centre for Forestry and Forest Products.

D. Noack,
Head, Federal Research Centre
for Forestry and Forest
Products,
Hamburg

OPENING REMARKS

Dear colleagues, ladies and gentlemen.
First of all I welcome you to the Institute of Forest Genetics and Forest Tree Breeding of the Federal Research Centre for Forestry and Forest Products. You came here to inform each other on the latest developments in 'Somatic Cell Genetics of Woody Plants', discuss problems of mutual interest and learn from the experiences of each other. We are especially glad and honoured that this meeting takes place in Grosshansdorf.

In addition to Biotechnology, there are active research projects in the following areas in our Institute:
- Ecological Genetics and Provenance Research
- Breeding
- Resistance Research
- Conservation of Gene Resources

Because of problems in vegetative propagation in the forest tree species, a small but a very active project group was set up under the guidance of Dr. Raj Ahuja here at the Institute. Till some years ago, we were propagating aspens in limited numbers by rather expensive and laborious methods using root-suckers and green-shoots. Our first clonal trials based on such methods needed some four years to be able to produce clonal material and to establish them. Todays procedure requires no more than two steps for micropropagation: proliferation of micro-shoots on one medium, and rooting of microshoots in a peat-perlite substrate, and total of one year for planting in the field.

But besides callus, organ, cell and protoplasts cultures for regenerating plants, we expect more from biotechnology, as for example:
- to detect useful variants for morphological, biochemical and genetic traits
- to establish haploid tissues for production of diploid homozygous lines and for their utility to produce specific genotypes for hybridization
- to preserve germplasm for future breeding and forestation programs
- to test and select early for resistance or tolerance against biotic and abiotic factors under *in vitro* conditions
- to fuse protoplasts to produce new genotypes that under natural conditions are not producible by sexual hybridization
- to incorporate foreign genetic material into plant cells by the help of plasmid-vector systems.

Most of the techniques might be directed at shortening the breeding cycles. However, we should make sure that application of biotechnology does not lead to reduction of genetic variability, that is so essential for the survival of forest tree populations.

I wish you all a successful and pleasant stay in Grosshansdorf.

G. H. Melchior
Director, Institute of Forest
Genetics & Forest Tree Breeding,
Grosshansdorf

INTRODUCTION

Most forest tree species were considered recalcitrant a decade ago, but now with the improved in vitro techniques some progress has been made towards culture of tree species. Micropropagation has been achieved from the juvenile tissues of a number of forest tree species. On the other hand, tissues from most mature trees are still very difficult to grow and differentiate in vitro. Nevertheless, there has been slow but steady progress in the application of tissue culture technology for culture of tissues, organs, cells and protoplasts of tree species.

As compared to most agricultural crops, and herbaceous plant species, trees are a different lot. They have long generation cycles. They are highly heterozygous and have a large reservoir of genetic variability. Because of this genetic variability, their response in vitro is also variable. On a single medium, the response of tissues from different trees (genotypes) of a single species may be quite different: some responding by induction of growth and differentiation, while others showing minimal or no growth at all. That makes the somatic cell genetics of woody plants somewhat difficult, but at the same time interesting.

The first Workshop of the International Union of Forestry Research Organizations (IUFRO) Working Party S2.04-07 'Somatic Cell Genetics' was held at Grosshansdorf, Federal Republic of Germany, 10-13 August, 1987. This proceeding is an outcome of that Workshop 'Somatic Cell Genetics of Woody Plants'. Recent advances in the somatic cell genetics of woody plants were discussed at this Workshop. There were six major sessions at this Workshop: somatic embryogenesis; genetic transformation; genetic control of morphogenesis; in vitro screening, somaclonal selection; and testing and deployment; and poster presentations. Some 40 participants from 10 countries of the world participated in the Workshop.

The Workshop provided a forum for an exchange of ideas and discussions in the area of somatic cell genetics and related areas of woody plants.

M. Raj Ahuja
Chairman,
IUFRO Working Party S2.04-07,
Somatic Cell Genetics

DEVELOPMENT AND CHARACTERIZATION OF *IN VITRO* EMBRYOGENIC SYSTEMS IN CONIFERS

BECWAR, M.R., WANN, S.R., JOHNSON, M.A., VERHAGEN, S.A., FEIRER, R.P., AND NAGMANI, R.

The Institute of Paper Chemistry, Appleton, WI 54912

1. ABSTRACT

Our progress is reviewed on development of somatic embryogenesis in conifers for mass propagation. A distinct embryogenic callus (EC) phenotype, white, mucilaginous, and rapidly growing, has been initiated on modified MS media with 2,4-D or NAA (2-5 mg/L) and BA(0-1 mg/L) from immature embryos of Norway spruce (*Picea abies*), white spruce (*Picea glauca*), loblolly pine (*Pinus taeda*), pond pine (*Pinus serotina*), and white pine (*Pinus strobus*). EC has also been initiated from mature embryos of Norway spruce and maintained as rapidly growing (48 hour doubling) liquid suspensions. Initiation of EC in *Picea* and *Pinus* differ markedly in several ways. Precotyledonary embryos were optimal in *Pinus* and EC originated from the suspensor region. In *Picea* EC originated from the hypocotyl and cotyledon region of predominantly post-cotyledonary embryos. Biochemically, EC of *Picea* and *Pinus* were similar and distinctly different from nonembryogenic callus (NEC) in terms of ethylene evolution rates (EC low and NEC high), level of total reductants, including glutathione (EC low and NEC high), and protein synthesis rates (EC high and NEC low). Conifer somatic embryos contained proplastids closely resembling those found in early zygotic embryos. On proliferation medium in the light, EC was white and maintained the proplastid morphology, whereas, NEC was green and contained mature chloroplasts with grana. These biochemical and ultrastructural differences served to both verify and predict embryogenic potential.

With Norway spruce somatic embryos, maturation frequencies as high as 25% have been attained. Germination frequencies as high as 82% (mean 56%) have been obtained. Twenty-nine percent of the somatic embryo plantlets survived transfer to the greenhouse, set a dormant terminal bud, overwintered to -5°C, and renewed vegetative growth synchronously with control seedlings. This is the first report of overwintering and renewed vegetative growth from resting buds of conifer somatic embryo plants.

2. INTRODUCTION

The long life cycle of conifers slows genetic improvement via the traditional sexual breeding process. Clonal (vegetative) propagation techniques show considerable promise for achieving more rapid tree improvement and increased productivity (7,15,20,29). There are three distinct methods of clonal propagation that are applicable to forest trees: macropropagation, micropropagation, and somatic embryogenesis. Macropropagation, the rooting of stem cuttings, is widely used in certain hardwoods as Eucalyptus spp. and softwoods as Norway spruce (*Picea abies*). Micropropagation involves regeneration from small pieces of tissue from either preexisting meristems, e.g., axillary buds, or from adventitious buds. Although micropropagation has proven utility in the clonal propagation of several commercially important forest trees, including Eucalyptus spp. and radiata pine (*Pinus radiata*), its primary limitation for mass propagation is the high produc-

M. R. Ahuja (ed.), Somatic Cell Genetics of Woody Plants, 1–18.

tion cost of individual propagules. Recent advances in regeneration via adventitious bud culture suggest that the technique may be economical for radiata pine (1).

A more promising clonal propagation technique for the economical production of large numbers of propagules is somatic embryogenesis, the production of embryolike structures from somatic tissue under *in vitro* conditions. Somatic embryos can be produced from cells, thus making highly efficient liquid cell culture techniques available for maintenance and production purposes. Furthermore, somatic embryos can be encapsulated to form artificial seeds for highly efficient delivery to existing tree nursery programs (27).

Somatic embryogenesis has been reported for several coniferous species, including Norway spruce (16), European larch (*Larix decidua*) (24), radiata pine (28), sugar pine (*Pinus lambertiana*) (13), loblolly pine (*Pinus taeda*) (14), and white and black spruce (*Picea glauca* and *P. mariana*) (19). In our laboratory we have initiated embryogenic cultures of Norway spruce (3), white spruce (25), loblolly pine, pond pine (*Pinus serotina*), and white pine (*Pinus strobus*) (5,35).

Here we review our progress on development and characterization of *in vitro* embryogenic systems in conifers. Emphasis is placed on similarities and differences between embryogenic systems in *Picea* and *Pinus*.

3. INITITATION OF EMBRYOGENIC CALLUS IN CONIFERS

3.1 *Picea*

3.1.1. Origin of embryogenic callus. Immature embryos of Norway spruce and white spruce produce both an embryogenic callus (EC) and a nonembryogenic callus (NEC) when cultured on basal medium supplemented with auxin and cytokinin (16,19). Recent investigations in our laboratory have provided detailed information on the nature and site of origin of the EC (25). Immature embryos of *Picea* cultured on callus induction medium produced two types of white to translucent calli that were phenotypically distinct. The callus that proliferated from the hypocotyl region was white to translucent, glossy, mucilaginous, and embryogenic. The epidermal and subepidermal tissue of the hypocotyl gave rise to the mucilaginous callus. After about 10 days in culture this callus gave rise to early stage somatic embryos consisting of an embryonal initial and suspensor initial cells. These results showed that in *Picea* a callus phase preceded somatic embryogenesis. The other type of white callus originated from the radicle and was nonembryogenic. Other investigations (19) have provided corroborative evidence for the hypocotyl origin of EC in *Picea*.

Numerous immature and mature embryos of *Picea* have been cultured on a variety of callus-inducing media, and initiation of EC from the radical region as reported by Gupta and Durzan (12) was not observed. We have observed initiation of EC at very low frequencies from cotyledons of 10-14 day old germinants of Norway spruce. This was reported earlier by Krogstrup (22). Recently, Lelu (23) initiated EC from 23% of cotyledons of 3-7 day old germinants of Norway spruce.

3.1.2. Optimum initiation window. In Norway spruce EC was initiated from immature embyros on the basal medium of von Arnold and Eriksson (31) supplemented with 2,4-D and BA (2 and 1 mg/L, respectively). All *Picea* cultures were initiated and maintained at 23°C with 16 hr irradiance (15-50 $\mu E \cdot m^{-2} \cdot sec^{-1}$) from cool-white fluorescent and incandescent lights. Initiation frequencies as high as 75% were attained from embryos collected in Wisconsin during July, 1985 (3). EC was also initiated from a different source tree during the summer of 1986 and the results are summarized in

Table 1. During the four week period from June 30 to July 21, initiation of EC was significantly higher than earlier or later collections. During the period of high initiation, embryo explants were predominantly cotyledonary. A 14-day cold (4°C) storage period prior to culture initiation did not significantly increase initiation (Table 1). Hakman and von Arnold (17) reported that a cold pretreatment of cones for two months increased EC initiation. The 14-day cold pretreatment period we used may not have been long enough to significantly increase initiation. Regardless, these results demonstrate the utility of cold storage for extending the time available for initiation from immature embryos of *Picea*.

TABLE 1. Frequency of initiation of EC from immature embryo explants of freshly collected and cold stored cones of Norway spruce.

Collection time month/date	Explant length (mm)	Explants with cotyledons, (%)	Embryogenic callus initiation (%) from explants[1]	
			fresh[2]	cold stored
6/23	0.2 ± 0.1	0	6 a	22 a
6/30	0.5 ± 0.1	0	59 b	34 b
7/7	0.9 ± 0.4	64	44 b	6 a
7/14	2.3 ± 0.4	83	62 b	72 b
7/21	3.4 ± 0.2	100	55 b	65 b
8/4	--	100	0 a	--

[1]Twenty-five explants per each treatment, 5 explants per plate. Data are mean values among plates. Means within rows followed by the same letter are not significantly different ($p = 0.05$).

[2]Means followed by the same letter are not significantly different.

The optimum window for initiation of EC from immature embryos of white spruce has also been determined (Figure 1). During the summer of 1986 a total of 388 immature embryo explants of white spruce were cultured on a modified von Arnold and Eriksson (31) basal medium supplemented with 2,4-D and BA (2 and 1 mg/L, respectively). Culture conditions were identical to those used for initiation from immature Norway spruce embryos (3). Cotyledonary embryos of about 1.5 to 2.0 mm were the most effective explants for initiation of EC. Thus, embryo length and the stage of development of cotyledonary primordia were useful indices for identifying the optimum stage for initiating EC from white spruce immature embryos.

The window for initiation of EC in Norway spruce has been extended to mature enbryos by utilizing a modified basal medium, BLG. This medium as described by Amerson (2), is a modified MS medium. The modifications include replacement of NH_4NO_3 with 10 mM glutamine; reduction of KNO_3 from 1900 to 100 mg/L, $MgSO_4$ from 370 to 320 mg/L, sucrose from 3 to 2%; and the addition of 100 mg/L asparagine and 745 mg/L of KCl. The results are fully summarized elsewhere (6). In brief, the results showed that by culturing mature embryos in the light (16 hr photoperiod) on half-strength BLG with either 2,4-D or NAA and BA, approximately 25% of the explants intitiated EC and a majority of the lines were successfully maintained. In contrast, initiation of EC from mature embryos was achieved at a very low frequency (< 3%) on the protocol used for immature embryos, and none of these lines could be maintained. Furthermore, the change to full-strength BLG did not result in initiation of EC from mature embyros. These results point out the importance of the levels and interactions of several components of the

BLG medium for extending the initiation of EC to mature embryos. Von Arnold and Hakman (32) have been able to initiate EC from mature embryos of Norway spruce cultured in the dark on the von Arnold and Eriksson (31) medium by reducing the sucrose level from 3.4 to 1%. Taken collectively, our results on initiation from mature embryos and those of von Arnold (30) demonstrate that changes in and optimization of medium components can significantly affect initiation and therefore play a major role in extending the initiation window to more mature tissues.

3.1.3. Quantification of embryogenic capacity. A method for quantitative determination of the level of somatic embryogenesis in conifer EC has been developed (4). EC of Norway spruce was dispersed in liquid by agitation and plated in a thin layer of medium containing 0.6% low melting point agarose. The density of somatic embryos ranged from 200 to 1500 per gram of EC among 11 lines surveyed. Thus, the somatic embryo counting technique was useful for identifying highly embryogenic lines among those with similar phenotypes. This technique has also proven useful for evaluating the effectiveness of biochemical treatments aimed at enhancing the level of embryogenesis (33).

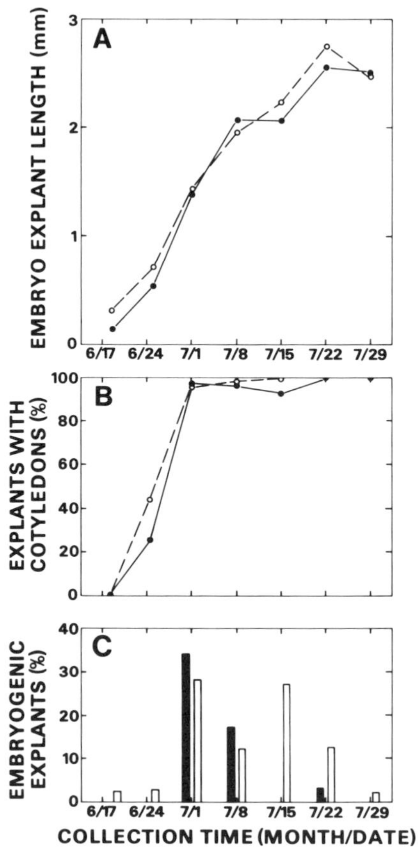

FIGURE 1. The stage of development of white spruce immature embryo explants: (A & B) and the frequency of initiation of EC(C). Solid symbols, tree 1 and open symbols, tree 2.

3.2. Pinus

Two culture protocols were used to initiate EC in *Pinus* spp. The first protocol used fertilized ovules as an explant, i.e., the female gametophyte with the intact embryo-suspensor complex. This technique was first reported effective for initiation of EC in radiata pine (28). The basal medium used, MSG, was a modified MS basal with NH_4NO_3 replaced by 1450 mg/L (10 mM) glutamine, the KNO_3 level reduced from 1900 to 100 mg/L, and the addition of 745 mg/L of KCl. The basal medium was supplemented with 1% activated charcoal. The second protocol used was the culture of isolated immature embryos. The basal media used were MSG and DCR1. DCR1 was a modified DCR medium used by Gupta and Durzan (11) with the glutamine level increased from 50 to 250 mg/L. Various levels of 2,4-D and BA were added to the basal media on which immature embryos were cultured. All cultures were grown in the dark at 23°C.

3.2.1. Pond pine. Figure 2A shows extruded callus from a loblolly pine ovule. Pond pine extruded callus was phenotypically similar. The frequency at which a white to translucent and mucilaginous callus was extruded from the archegonial end of ovules of pond pine is shown in Table 2. Although up to 12% of the ovules initiated the extruded callus phenotype, only some of these cultures could be maintained (Table 2). Initiation, as used here, refers to initial formation of the extruded callus, whereas maintenance refers to lines established in culture for over one year. To determine the origin of the extruded callus, ovules were cut open when the callus was removed from the ovule for subculture. In all cases the primary embryo had neither developed or atrophied to a significant degree, and callus proliferation was confined to the suspensor region. Histological examination of the extruded callus revealed a mixture of unaggregated suspensor-like cells and globular clumps of densely cytoplasmic cell which resembled pre-embryonal masses (Figure 2B). The majority of the globular structures appeared similar to somatic embryos lacking suspensors. Further histological examination of the extruded callus revealed the presence of early stage somatic embryos, containing an embryonal mass and attached suspensor-like cells (Figure 2C).

TABLE 2. Initiation and maintenance of mucilaginous callus extruded from ovules of pond pine.

Collection time (month/date)	Germination[1] (%)	Extruded callus: Initiation[2] frequency (%)	Extruded callus: Lines maintained
7/11	0	12	0
7/18	8	2	2
7/25	68	5	1
8/1	96	2	1
8/8	--	1	0

[1]Germination efficiency of embryos cultured on basal medium.
[2]One-hundred ovules cultured per collection.

Smith (28) reported that the frequency at which the extruded callus was formed was dependent on the stage of embryo development. Specifically, his

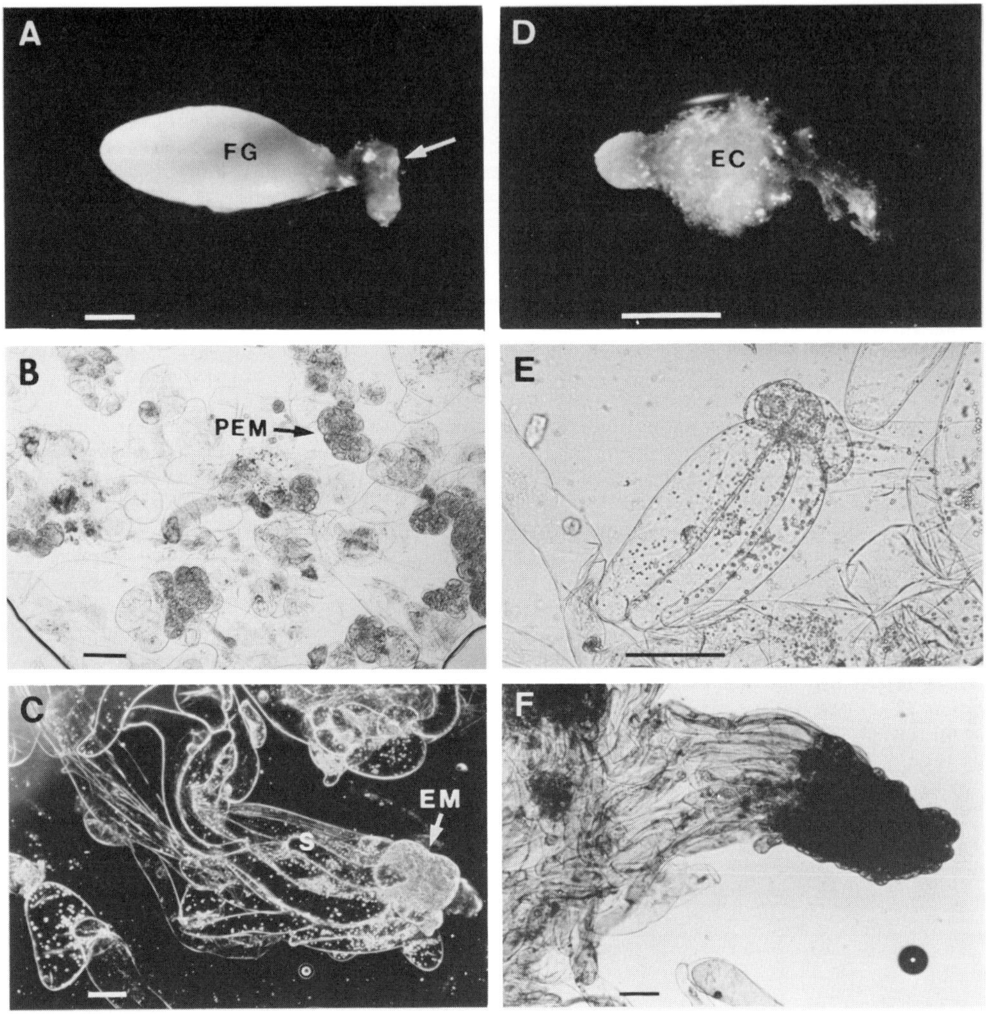

FIGURE 2. Initiation of somatic embryogenesis in Pinus. A: White-mucilaginous callus extruded (arrow) from female gametophyte (FG) of loblolly pine. B: Micrograph of extruded callus of pond pine showing pre-embryonal masses (PEM). C: Early stage somatic embryos of pond pine showing embryonal mass (EM) and attached suspensor-like cell(S). D: Origin of embryogenic callus (EC) from suspensor region of loblolly pine embryo. E: Early stage somatic embryo of loblolly pine. F: White pine somatic embryo. Scale bars: 1 mm in A and D; 100 μm in B, C, E, and F.

results showed that as embryos matured within the ovule (and embryo germination was possible) the initiation frequency dropped off. In pond pine the extruded callus was initiated from both ovules of early collections which contained predominantly pregerminable embryos and ovules of collections as late as August 1, which contained germinable embryos (Table 2).

3.2.2. Loblolly pine. The time course of embyro development in loblolly pine during the period in which cultures were initiated is shown in Figure 3. Note that embryo explants cultured on July 14 to 28 were predominantly precotyledonary and had a mean length of 1 mm or less, whereas embryos cultured during August were mostly cotyledonary and 1 mm or larger.

The stage of explant development was one of the most critical factors for successful initiation of EC from immature embryos of *Pinus*. The results of our initial experiments (Table 3) showed that EC was only established from precotyledonary embryos which were less than 0.3 mm in length. Subsequent experiments have verified that the precotyledonary immature embryo stage is optimum for initiation and establishment of EC. Initial results also showed that low levels of 2,4-D were effective for initiation of EC, whereas higher levels (10 mg/L) were ineffective (Table 3). The EC originated from the suspensor region of the loblolly pine embryos (Figure 2D), in agreement with the site of origin of EC reported for sugar pine (13). Histologically, the EC callus initiated from immature embryos of loblolly pine was similar to the extruded EC callus of pond pine, i.e., a heterogeneous mixture of unaggregated suspensor-like cells, globular clumps of highly cytoplasmic cells, and early stage somatic embryos (Figure 2E).

3.2.3. White pine. The time course of embryo development during culture initiation in white pine is shown in Figure 4. Note that embryo explants derived from the first two collections of cones (July 2 and 9) were mostly precotyledonary and less than 1 mm in length, whereas those from the last three collections were mostly cotyledonary and greater than 2 mm.

The results in Table 4 show that precotyledonary embryos of white pine (e.g., embryos from the first two collections) were effective explants for initiation of EC, whereas cotyledonary embryos were ineffective. The EC originated from the suspensor region of the immature embryo explants. Thus, initiation of EC in white pine was similar to loblolly with respect to optimal stage of explant development and site of origin. Although EC was initiated on both MSG and DCR basal medium supplemented with 2,4-D and BA (Table 4), no EC was initiated on the DCR 3/0 treatment (data not shown), suggesting a cytokinin requirement for initiation. A white pine somatic embryo is shown in Figure 2F.

3.3. Comparison of *in vitro* embryogenesis in *Pinus* and *Picea*

A visual inspection of EC in *Pinus* and *Picea* suggests phenotypic similarity. They are both white to translucent and mucilaginous. Further comparative observations of *Pinus* and *Picea* EC show marked differences in initiation and growth characteristics.

The most striking differences in initiation include differences in the optimum stage of explant development, the site of origin, and the frequency of initiation of EC. In *Picea* postcotyledonary embryos were optimum and the EC originated from the hypocotyl (25) and cotyledon regions (22,23). In *Pinus* precotyledonary embryos were optimum and EC originated from the suspensor region. Initiation frequencies were relatively high in *Picea*, 40 to 75% from immature embryos and 25% from mature embyros. In contrast, initiation frequencies were consistently low in *Pinus*, typically less than 3%. It is possible that changes in medium components may increase initiation frequencies in *Pinus* and enable initiation from more mature tissues. As indicated previously, media modifications were effective in extending the window to mature embryos in Norway spruce. Another difference between Pinus and Picea was that auxin and cytokinin were obligatory for initiation in Picea but were not required for initiation of EC from fertilized ovules of pond pine.

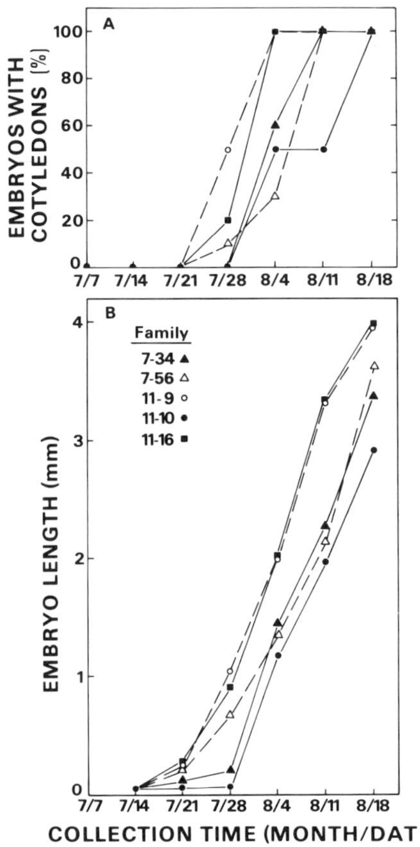

FIGURE 3. The stage of development of immature embryo explants of five families of loblolly pine. A: Percentage of embryos with cotyledonary primordia. B: Embryo length.

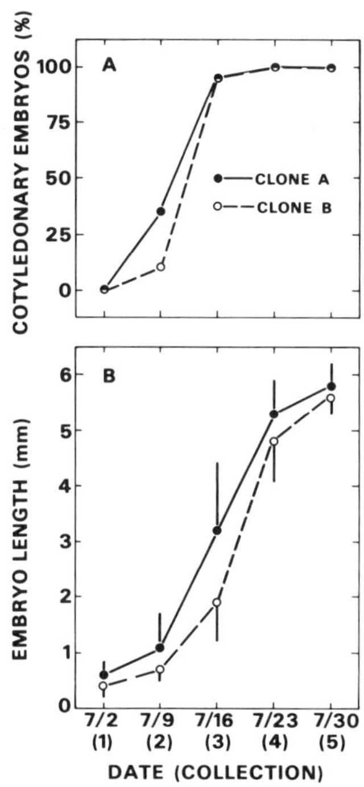

FIGURE 4. The stage of development of immature embryo explants of white pine collected during July 1986 in Wooster, OH. A: Percentage of embryos with cotyledonary primordia. B: Embryo length.

Proliferative growth of EC also differs considerably between Pinus and Picea. In Pinus spp., EC was often dominated by pre-embryonal masses (see Figure 2B) or very early stage somatic embryos (see Figure 2C and E). However, in Picea spp. somatic embryos reached a more advanced stage of development on proliferation medium.

TABLE 3. Effect of the stage of explant development and media modifications on establishment of EC from immature embryo explants of loblolly pine.

Stage of explant development[1]			Frequency (%) of embyrogenic lines maintained			
			Basal medium (A/C mg/L)[3]			
embryo size (mm)	embryos with cotyledons (%)	N[2]	DCR 10/.5	DCR 3/.5	MSG 10/1	MSG 2/1
0.1-0.3	0	130	0	2.3	0	1.5
0.7-1.2	37	85	0	0	0	0

[1]Embryo explants derived from cones collected July 21 to Aug. 4, 1986 from five families of loblolly pine. See Figure 3.
[2]N = number of explants cultured per each medium treatment.
[3]Refer to text for description of basal medium. A/C refers to auxin and cytokinin levels.

TABLE 4. Effect of the stage of embryo development on initiation of EC in white pine.

Collection (month/date)	Embryos with Cotyledons (%)	N[2]	Embryogenic callus[1] initiation frequency (% explants) Medium MSG 2/1	DCR 3/.5
1 (7/2)	0	150	--	2.7
2 (7/9)	23	230	1.3	2.2
3 (7/16)	95	150	0	0

[1]Initiation from cones stored at 4°C for 2-3 months. Cones were derived from five trees: clones A and B (Figure 4) collected in Wooster, OH and three from Freedom, WI.
[2]N = number of explants cultured per each medium treatment.

4. BIOCHEMICAL CHARACTERIZATION OF EMBRYOGENIC CONIFER CALLUS

Biochemical analyses of Picea and Pinus EC and NEC were conducted as previously described (33-35). Briefly, ethylene was determined by gas chromatography, protein synthesis was measured by 3H-leucine uptake, glutathione (GSH) was measured by the rereduction of oxidized GSH by commercial GSH reductase, and the level of total reductants was measured by the ability to reduce potassium ferricyanide. In Picea spp., EC and NEC assayed were derived from the same genotype (explant) and were cultured under identical conditions. In Pinus spp., although EC and NEC were initiated and maintained under similar or identical conditions, the callus types were derived from different embryo explants, and were therefore genotypically different.

Significant differences in biochemical parameters were detected between EC and NEC in both Pinus and Picea spp. (Table 5). These differences were of sufficient magnitude to suggest that they are indicative of an embryogenic condition in conifers. Although interspecific differences were detected, relative to NEC, EC 1) evolved less ethylene, 2) contained lower amounts of GSH and other nonspecific reducing agents, and 3) synthesized or "turned over" protein at a faster rate. Similar trends in protein synthesis rates and level of total reductants between EC and NEC of European larch have also been observed (data not shown). These results suggest that EC callus in conifers is biochemically and metabolically similar irrespective of species, and distinctly different from NEC. While these conclusions are broadly stated for conifers, white spruce stood out as not exhibiting clear biochemical differences between EC and NEC. This may in part be due to the fact that when grown under proliferative conditions, somatic embryos of white spruce can contain anthocyanins which would reduce potassium ferricyanide.

TABLE 5. Biochemical differences between EC and NEC of conifers.

	Ratio of measured parameter (NEC/EC)[1]			
Species	Protein synthesis rate	Total reductants	GSH	Ethylene evolution rate
loblolly pine	0.03	4	9	345
white pine	0.02	13	3	127
pond pine	0.4	10	8	14
Norway spruce	0.03	17	3	12
white spruce	0.4	2	1.2	2.4

[1]Only white spruce ratios not significantly different ($p = 0.05$) from 1.0.

The biochemical assays performed were all rapid, convenient, and required small amounts of tissue. In the case of ethylene, the assay was nondestructive, enabling reuse of the tissue in other experiments. These attributes make the assays attractive candidates for markers of embryogenic potential. However, biochemical markers may have limited utility in conifers because the EC phenotype can be easily recognized. It seems that biochemical assays of EC will be of greater utility in identifying key metabolic pathways involved in growth and development of somatic embryos into plants.

5. ULTRASTRUCTURAL CHARACTERIZATION OF EMBRYOGENIC CONIFER CALLUS

The chloroplasts of light grown conifers are typical of those found in most higher plants. The organelles, usually 2-10 μm in length, contain starch grains, protein bodies and internal thylakoid membranes organized into grana. The organization of the thylakoid membranes is dependent, however, upon the age and physiological state of the tissue.

In contrast, plastids in EC of spruce had a unique morphology at the ultrastructural level (Figure 5A & B). The plastids, appeared more darkly stained (more electron dense) than mitochondria, lacked the internal organization of a mature chloroplast. Some of these plastids contained small starch grains, although they were not nearly as large as the starch grains present in mature chloroplasts in leaf or cotyledon tissue. The plastids present in the green NEC of spruce, on the other hand, appeared similar to a typical chloroplast (Figure 5C & D). Thylakoid membranes, some organized into grana, were found in all the chloroplasts. Large starch grains were also prominent in some of the chloroplasts.

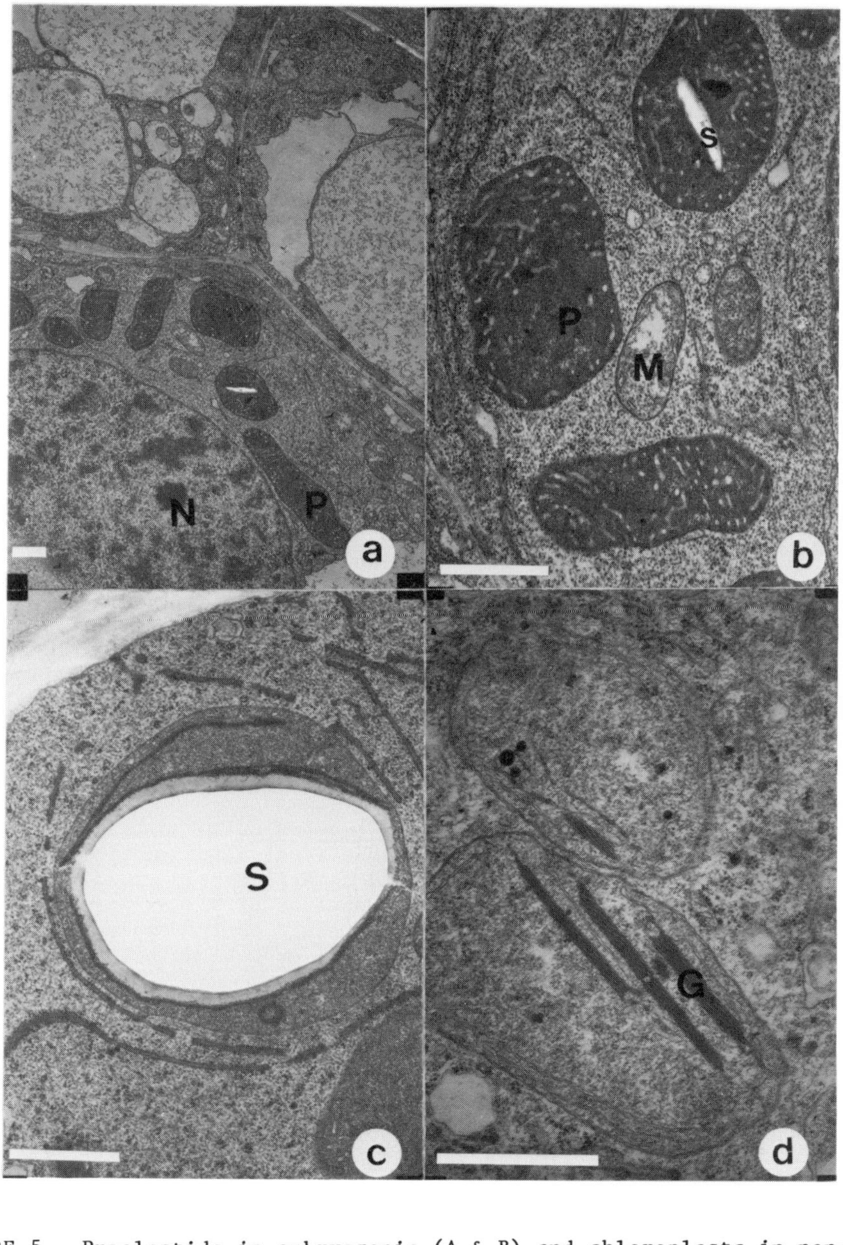

FIGURE 5. Proplastids in embryogenic (A & B) and chloroplasts in non-embryogenic (C & D) callus of Norway spruce. P = proplastid. M = mitochondria. S = starch grain. G = grana. N = nucleus. Scale bars = 1 μm.

Zygotic embryos excised from mature Norway spruce seeds were also examined. The plastids resembled those observed in EC of spruce. That is, they also lacked organized thylakoid membranes (photographs not shown). In order to determine if the plastid morphology observed in zygotic and somatic embryos was unique to Norway spruce, EC of loblolly pine, pond pine, white pine, and European larch were examined (10). EC of these species all contained plastids that closely resembled those in EC of spruce (photographs not shown). In addition, the plastids observed in somatic embryos of carrot exhibited the same morphology as those in EC of conifers.

Further investigations showed that as Norway spruce somatic embryos matured (cotyledons begin to form and green) chloroplasts with thylakoids and grana became evident. We have observed the same phenomenon in carrot, where somatic embryos beyond the torpedo stage, if grown in the light, contained green mature chloroplasts. Thus, it appears that proplastids are indicative of early stages of embryonic development, whether *in vivo* or *in vitro*. These findings provide evidence in support of our previous biochemical studies (9,33,34) that many aspects of somatic embryogenesis mimic or correspond to *in vivo* embryogenesis (21). In summary, chloroplast morphology may be one of the best indicators that embryogenesis in *Pinus* and *Picea* is occurring and proceeding "normally" *in vitro*.

6. CONIFER EMBRYOGENIC SUSPENSION CULTURES

For the purposes of mass propagation, a liquid embryogenic culture system is desirable for several reasons, including the following. Liquid cultures are easier and more economical to maintain than callus cultures. They have potential for higher growth rates (decreased cell doubling times), thus eliminating the need for maintaining large stocks of callus. Lastly, liquid culture systems are amenable to automation of both somatic embryo proliferation and development *en masse* (e.g., large bioreactor culture systems). In addition to their utility for mass propagation, liquid cultures serve as an ideal source for cells and protoplasts for genetic modification experiments.

Although establishment of embryogenic suspension cultures has been rereported in both *Picea* and *Pinus* (14,18), quantitative information has not been presented on growth characteristics and somatic embryo yield. We have established rapidly growing embryogenic suspensions of Norway spruce. The cultures were derived from EC initiated from mature embryo explants on half-strength BLG with 2 mg/L 2,4-D or NAA and 1 mg/L BA as previously described.

During the linear phase of growth, the liquid cultures had a doubling time of about 48 hours and reached a somatic embryo density of about 100/mL (Figure 6). Rapid growth rates were maintained by repeated subculture at 10-14 day intervals (1:10 dilution with fresh medium).

With both high growth rates and somatic embryo yield, the Norway spruce suspension cultures show promise as an ideal system for mass production of somatic embryos. Continued efforts are directed at obtaining maturation of the somatic embryos while still in the liquid culture system.

7. DEVELOPMENT OF CONIFER SOMATIC EMBRYOS TO PLANTS

The Norway spruce somatic embryos can be converted to phenotypically normal plants (Figure 7). Somatic and seedling plants overwintered and renewed vegetative growth synchronously. A complete description of the frequency of plant regeneration is presented elsewhere (6). Based on optimum maturation frequencies (25%), germination (56%), and conversion to plants (29%), the highest mean efficiency of 4% was attained. Because of

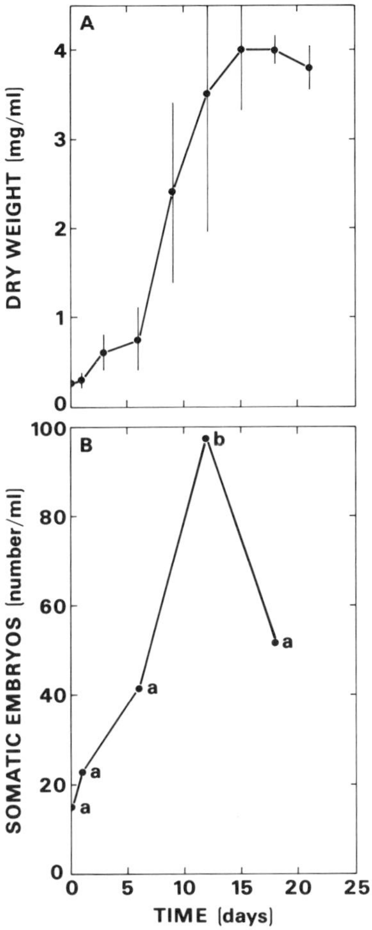

FIGURE 6. Growth curve (A) and somatic embryo density (B) of liquid suspension cultures of Norway spruce grown on half-strength BLG with 2 mg/L NAA and 1 mg/L BA. Vertical bars in A are ± standard deviation among triplicate samples. Means of triplicate counts in B followed by unlike letters are significantly different (p = 0.05).

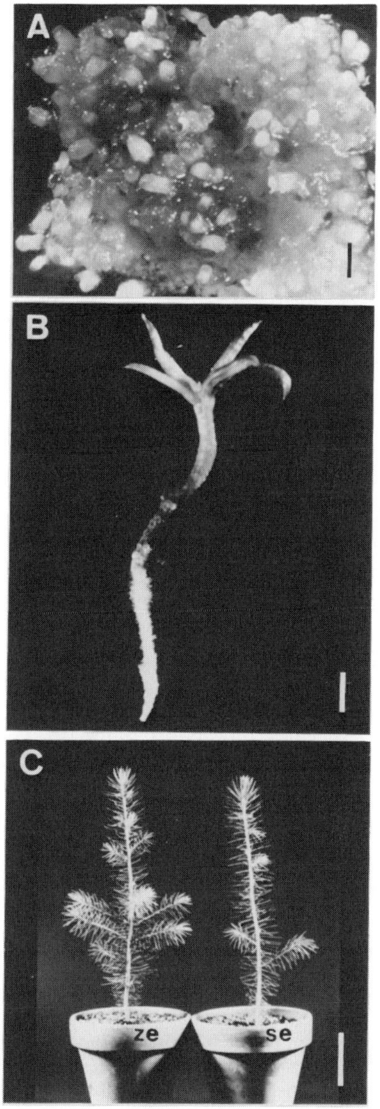

FIGURE 7. Maturation (A), "germination" (B), and conversion (C) of Norway spruce somatic embryos to plants. C: Comparison of somatic embryo plant (se) and control plant grown from zygotic embryo (ze). Scale bars: 1 mm in A & B; 5 cm in C.

the exceedingly long regeneration cycles of conifers, it is of prime importance to verify at an early stage that the somatic plants are uniform and genetically "true-to-type." Current efforts are directed toward production of large numbers of Norway spruce somatic plantlets for quantitative analysis of uniformity and growth characteristics.

8. COMPARISON OF *IN VIVO* AND *IN VITRO* CONIFER EMBRYOGENESIS

Polyembryony, the formation of multiple embryos, is common *in vivo* in conifers. There are two types of polyembryony, simple and cleavage (8,26) (Figure 8). Simple polyembryony, the fertilization of more than one egg per female gametophyte, occurs in both *Picea* and *Pinus*. Therefore, the resulting embryos may be genotypically different. In addition to simple polyembryony, cleavage polyembryony occurs in *Pinus* (Figure 8). Cleavage embryos result from the division of apical tier cells of an individual proembryo and are, therefore, genetically identical.

Observations of *in vitro* embryogenesis in both *Picea* and *Pinus* have suggested that one mechanism of embryo formation may be a cleavage type of process. That is, multiple embryos were found with common suspensors. Although further investigations are needed to fully characterize possible types of *in vitro* embryogenesis in conifers, the following suggest the possibility of differences between *Picea* and *Pinus*. As indicated previously, in *Picea* a callus phase preceded initiation of somatic embryos (25). Therefore, in *Picea* besides cleavage formation of embryos *in vitro*, another type of embryo initiation mechanism is operative. In *Pinus* initiation of embryogenesis *in vitro* may simply be a reinitiation of the cleavage process (which occurs *in vivo*).

9. SUMMARY

Much progress has been made on the development of *in vitro* embryogenic systems in conifers since the first reports of somatic embryogenesis in Norway spruce (16) and European larch (24) in 1985. The embryogenic callus phenotype, white to translucent and mucilaginous, has been similar in all conifers. Furthermore, our biochemical studies have provided evidence of the physiological similarity of embryogenic conifer callus among species. Although similar physiologically, EC in *Picea* and *Pinus* have different optimum initiation windows and sites of origin from immature embryo explants. These differences may relate to inherent differences in embryology between *Picea* and *Pinus*.

Considerable progress has also been made on regeneration of plants from conifer somatic embryos. Several reports have verified that conifer somatic embryos reached the germination stage (3,12-14,17,19,24,30). Our results (6) indicate that Norway spruce somatic embryo plants were similar to seedling derived plants in terms of physiological response to changing environment. This is the first demonstration of overwintering and renewed vegetative growth from resting buds of conifer somatic embryo plants.

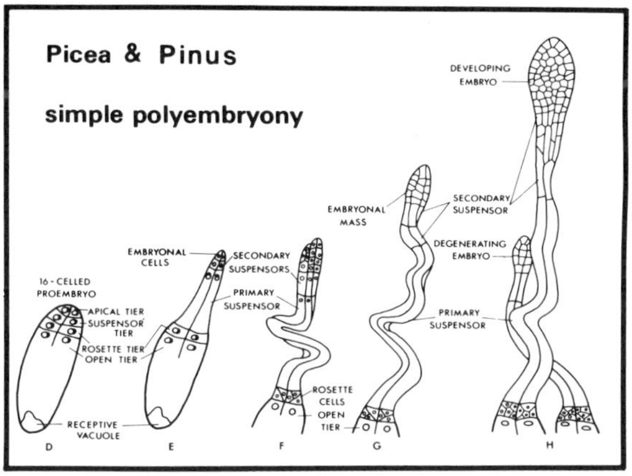

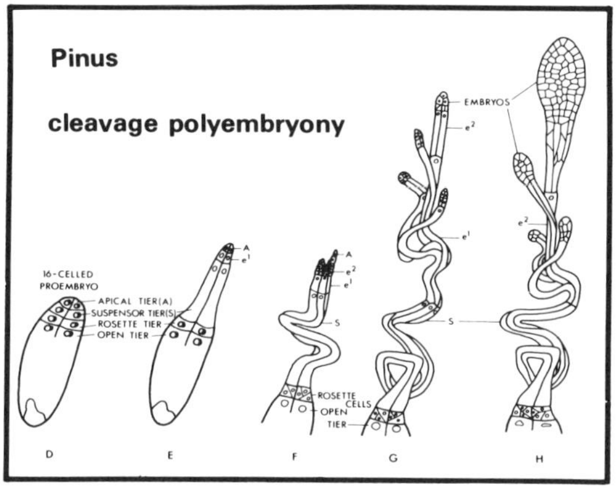

FIGURE 8. *In vivo* embryogenesis in *Picea* and *Pinus*. Used with permission from Owens and Blake (26).

10. ACKNOWLEDGMENTS

We wish to acknowledge all members of the IPC Forest Biology Division who participated in this research. In particular, Debbie Hanson, Lynn Kroll, Judy Wyckoff, John Carlson, Julian Conkey, Gary Wyckoff, Robert Arvey, and Egon Hummenberger. We also thank the following individuals for supplying conifer cones: Lee Handley (Westvaco Corp., Summerville, SC), George Lowertz (Union Camp Corp., Rincon, GA), and Howard Kriebel (Ohio St. Univ., Wooster, OH).

11. REFERENCES

1. Aitken-Christie J, Singh AP, and Davies H: A new tissue culture system for radiata pine. In: Genetic Manipulation of Woody Plants. Eds., Hanover JH and Keathley DE. Plenum Press, NY. 1987.

2. Amerson HV, Frampton LJ, McKeand SE, Mott RL, and Weir RJ: Loblolly pine tissue culture: laboratory, greenhouse, and field studies. In: Tissue Culture in Forestry and Agriculture. Eds., Henke, R.R., _et al_. Plenum Press, NY. pp. 271-287, 1985.

3. Becwar MR, Noland TL, and Wann SR: Somatic embryo development and plant regeneration from embryogenic Norway spruce callus. Tappi J. **70**(4):155-160, 1987.

4. Becwar MR, Noland TL, and Wann SR: A method for quantification of the level of somatic embryogenesis among Norway spruce callus lines. Plant Cell Reports **6**:35-38, 1987.

5. Becwar MR, Wann SR, and Kriebel HB: Initiation of embryogenic callus in _Pinus strobus_ (eastern white pine) from immature embyro explants (abstract). In: Genetic Manipulation of Woody Plants. Eds., Hanover JH and Keathley DE. Plenum Press, NY. 1987.

6. Becwar MR, Verhagen SA, and Wann SR: The frequency of plant regeneration from Norway spruce somatic embryos. Proceedings, 19th Southern Forest Tree Improvement Conf., College Station, TX, June 16-18. The National Technical Information Service, Springfield, VA. 1987.

7. Bonga JM: Tree tissue culture applications. _In_: Advances in cell Culture, Vol. 5. Ed., Maramorosch K. Academic Press, Inc., NY, p. 209-39, 1987.

8. Buchholz JT. Origins of cleavage polyembryony in conifers. Bot. Gaz. **81**:55-71, 1926.

9. Feirer RP, Wann SR, and Einspahr DW: The effects of spermidine synthesis inhibitors on _in vitro_ plant development. Plant Growth Regul. **3**: 319-327, 1987.

10. Feirer RP, Wann SR, Becwar MR, Nagmani R, and Carlson J: A comparison of embryogenic calli: chloroplasts ultrastructure and protein synthesis (abstract). In: Genetic Manipulation of Woody Plants. Eds., Hanover JH and Keathley DE. Plenum Press, NY. 1987.

11. Gupta PK and Durzan DJ: Shoot multiplication from mature trees of Douglas-fir (_Pseudotsuga menziesii_) and sugar pine (_Pinus lambertiana_). Plant Cell Reports. **4**:177-179, 1985.

12. Gupta PK and Durzan DJ: Plantlet regeneration via somatic embryogenesis from subcultured callus of mature embryos of _Picea abies_ (Norway spruce). _In Vitro_. **22**:685-688, 1986.

13. Gupta PK and Durzan DJ: Somatic polyembryogenesis from callus of mature sugar pine embryos. Bio/Tech. **4**:643-645, 1986.

14. Gupta PK and Durzan DJ: Biotechnology of somatic polyembryogenesis and plantlet regeneration in loblolly pine. Bio/Tech. **5**:147-151, 1987.

15. Haissig BE, Nelson ND, and Kidd GH: Trends in the use of tissue culture in forest improvement. Bio/Tech. **5**:52-57, 1987.

16. Hakman I, Fowke LC, von Arnold S, and Eriksson T: The development of somatic embryos in tissue cultures initiated from immature embryos of Picea abies (Norway spruce). Plant Sci. **38**:53-59, 1985.

17. Hakman I and von Arnold S: Plantlet regeneration through somatic embryogenesis in Picea abies (Norway spruce). J. Plant Physiol. **121**: 149-158, 1985.

18. Hakman I and Fowke LC: An embryogenic cell suspension culture of Picea glauca (white spruce). Plant Cell Repts. **6**:20-22, 1987.

19. Hakman I and Fowke LC: Somatic embryogenesis in Picea glauca (white spruce) and P. mariana (black spruce). Can. J. Bot. **65**:656-659,1987.

20. Hasnain S, Pigeon R, and Overend RP: Economic analysis of the use of tissue culture for rapid forest improvement. The Forestry Chronicle, **6**(4):240-245, Aug., 1986.

21. Johnson MA, Carlson JA, Conkey JH, and Noland TL: Biochemical changes associated with zygotic pine embryo development. J. Expt. Bot. **38**: 518-524, 1987.

22. Krogstrup P: Embryo-like structures from cotyledons and ripe embryos of Norway spruce (Picea abies). Can. J. For. Res. **16**:664-668, 1986.

23. Lelu M, Boulay M, and Arnaud Y: Formation of embryogenic calli from cotyledons of Picea abies collected from 3 to 7 day old seedlings. C.R. Acad. Sci. Paris. **305**(series III):105-109, 1987.

24. Nagmani R and Bonga JM: Embryogenesis in subcultured callus of Larix decidua. Can. J. For. Res. **15**:1088-1091, 1985.

25. Nagmani R, Becwar MR, and Wann SR: Single-cell origin and development of somatic embryos in Picea abies (Norway spruce) and P. glauca (white spruce). Plant Cell Reports **6**:157-159, 1987.

26. Owens JN and Blake MD: Forest tree seed production. A review of the literature and recommendations for future research. Information report PI-X-53. Petawawa Nat. Forestry Inst., Can. For. Serv., Chalk River, Ontario. 161 pp, 1985.

27. Redenbaugh K, Paasch BD, Nichol JW, Kossler ME, Viss PR, and Walker KA: Somatic seeds: Encapsulation of asexual plant embryos. Bio/Tech. **4**: 797-801, 1986.

28. Smith DR, Singh AP, and Wilton L: Zygotic embyros of Pinus radiata in vivo and in vitro. Int. Conifer Tissue Culture Work Group. Abstracts. Ed. Smith, D.R., New Zealand Forest Service, Rotorua, New Zealand, 1985.

29. Timmis R, Abo El-Nil MM, and Stonecypher RW: Potential genetic gain through tissue culture. In: Cell and tissue culture in Forestry.

Eds., Bonga JM and Durzan DJ. Martinus Nijhoff Publ. Dordrecht, The Netherlands. Vol. 1, p. 198-215, 1987.

30. von Arnold S: Improved efficiency of somatic embryogenesis in mature embryos of *Picea abies* (L.) Karst. J. Plant Physiol. **128**:233-244, 1987.

31. von Arnold S and Eriksson T: *In vitro* studies of adventitious shoot formation in *Pinus contorta*. Can. J. Bot. **59**:870-874, 1981.

32. von Arnold S and Hakman I: Effect of sucrose on initiation of embryogenic callus from mature zygotic embryos of *Picea abies* (Norway spruce). J. Plant Physiol. **122**:261-265, 1986.

33. Wann SR, Feirer RP, Johnson MA, and Noland TL: Norway spruce as a model system for somatic embryogenesis in conifers. Proc. TAPPI Res. and Devel. Conf., Forest Biotech. Raleigh, NC, Sept. 28-Oct. 1, TAPPI Press, Atlanta, GA. pp. 131-135, 1986.

34. Wann SR, Johnson MA, Noland TL, and Carlson JA: Biochemical differences between embryogenic and nonembryogenic callus of *Picea abies*. Plant Cell Reports **6**:39-42, 1987.

35. Wann SR, Johnson MA, Feirer RP, Becwar MR, and Nagmani R: Biochemical differences between embryogenic and nonembryogenic callus of conifers (abstract). In: Genetic Manipulation of Woody Plants. Eds., Hanover JH and Keathley DE. Plenum Press, NY. 1987.

STUDIES ON EMBRYOGENESIS OF WOODY PLANTS IN CHINA

CHEN ZHENGHUA, YAO YUGUANG AND ZHANG LIHUA
Institute of Genetics, Chinese Academy of Sciences,
Beijing, China

0. ABSTRACT

During the last decade induction of somatic embryogenesis has been carried out in a number of plant species in China. Several features of embryogenesis of woody plants are described. The embryogenesis should go through two stages, that is, the young and the mature embryo stage, before they can be transferred to plantlet forming medium. Some factors affecting the induction frequency of embryogenesis, such as light and dark, temperature, medium and hormones are discussed. The causes of abnormal embryoid formation are also mentioned.

1. INTRODUCTION

Over the past decade much progress has been made in *in vitro* embryogenesis of woody plants in China. Plantlets have been produced from pollen in anther cultures of ten genera. With eight of these genera, i.e., *Hevea*, *Citrus*, *Vitis*, *Malus*, *Litchi*, *Euphoria* i.e. *Dimocarpus* Lour. is a kind of Chinese fruit tree, *Lycium* and *Morus*, the plantlets were obtained through embryogenesis (8,18), with two, i.e., *Populus* and *Camellia*, through organogenesis (8). For sixteen genera, i.e., *Populus* (30), *Eucalyptus* (22), *Actinidia* (16), *Elaeis* (9), *Elaeagnus* (23), *Crataegus* (24), *Vitis* (1), *Zizyphus* (2), *Prunus* (19), *Citrus* (25), *Camellia* (28), *Thea* (29), *Eriobotrya* (4), *Euphoria* (26), *Rhododendron* and *Mycetia* (Lu YL, personal communication), embryoids and plantlets were produced from somatic rather than from haploid tissues.

For mass cloning, embryogenesis is preferred over organogenesis because it produces embryoids rather than shoots that have to be rooted. Thus, one difficult step in the plantlet production procedure is bypassed. In addition, with further technical improvement it may eventually become possible to encapsulate somatic embryos, thus turning them into artificial seeds. This would further improve the efficiency of systems used for the mass production of plants from pollen and somatic tissues.

2. FACTORS AFFECTING EMBRYOID FORMATION

2.1. *Pretreatment*

Exposure of Chinese wolfberry anthers to low temperatures (3-5°C) for 120 hours before they were placed on the culture medium raised the frequency of embryoid induction from pollen three times (15). The pollen were at the mid-uninuclear stage during this low temperature treatment. Prior to the cold treatment the anther cells surrounding the pollen contained a large number of starch grains. After exposure to the low temperature this starch disappeared indicating increased amylase activity, and consequently increased availability of sugars for the

M. R. Ahuja (ed.), Somatic Cell Genetics of Woody Plants, 19–28.

dedifferentiation of the pollen. Simultaneously the cytoplasm became more dense. This may have promoted abnormal division of the pollen and thus initiated embryogenesis.

Exposing anthers of the rubber tree to low temperatures (6-11°C) for 24 hours caused disintegration of the microspores, but not of the cells of the anther wall. Therefore, cold treatment resulted in stimulation of embryogenesis from the somatic anther wall cells and in a strong reduction or absence of embryogenesis in pollen (6).

2.2. Explant

Of the sixteen genera in which embryoids and plants were produced from somatic cells (Table 1), nine formed embryoids if young embryos were used as explant. In four genera, endosperm explants provided embryogenic cultures; in two genera embryogenesis was obtained with stem segments or buds taken from 3-year-old plants raised from pollen. For obvious reasons one would prefer to induce embryogenesis in explants from mature, elite trees. However, to date this has only been possible with mature trees of *Mycetia*, young leaves of which have produced embryoids.

2.3. Media and its supplements

Since the media used to obtain embryogenesis in pollen have been reviewed recently (8), only a few details will be given here (Table 2). Murashige Skoog medium (MS) (20) has been the most often used medium for embryogenesis in both the pollen and somatic tissue cultures. This medium has a high level of inorganic salts and ammonium. Other media used were Gamborg B5 (14), Heller (H) (21) and ER (10).

In carrot cultures (17) an initial addition and subsequent withdrawal of 2,4-dichlorophenoxyacetic acid (2,4-D) is required to obtain embryoids. To induce embryogenesis in cultures of the tree species discussed here a cytokinin, i.e., N6-benzyladenine (BA), zeatin (ZEA) or kinetin (KIN), has to be added in addition to 2,4-D. After induction the 2,4-D is replaced by a weak auxin, i.e., naphthaleneacetic acid (NAA), indoleacetic acid (IAA) or indolebutyric acid (IBA), but cytokinin is retained.

Several of the cultures required casein hydrolysate (CH) or lactoprotein hydrolysate (LH) for proper embryogenesis.

2.4. Environmental factors

Anthers of seven grape varieties failed to form embryoids in the light, but an embryoid induction frequency of 0.6-9.0 % was achieved if they were cultured in the dark (31). By keeping suspension cultures of somatic cells of grape for one week in the dark before transfer to light, the embryoid induction rate was increased significantly (1). On the other hand, the frequency of induction of embryoids from pollen decreased when anthers of the rubber tree were cultured in the dark. Furthermore, while the roots of these embryoids developed properly, the growth of the shoots was strongly suppressed. Culture in diffuse light (12 hours/day) gave more normal development.

For all cultures discussed here a temperature range between 22 and 28°C was optimal for embryogenesis (6).

3. ORIGIN OF THE EMBRYOIDS

3.1. Embryoids from pollen

Depending on species, embryoid development followed a number of different pathways:

1) Embryoids develop from single calli, each of which has originated from an individual pollen grain.

2) A pollen grain develops one embryoid without prior callus formation. This embryoid soon forms a large number of secondary embryoids.

3) Individual pollen grains each form one, normally maturing embryoid.

Plantlets derived from pollen of *Hevea* (5), *Vitis* (31), *Litchi* (13), *Euphoria* (27), and *Malus* (12) were formed mainly via pathway 1 and 2, while those of *Lycium* (11), *Citrus* (3) and *Morus* (18) had followed 2 and 3. When pathway 3 is followed, each embryoid arising from one anther will have a different genotype. The reason for this is that the pollen in the anther have, because of meiotic segregation, differing genotypes. When pathway 1 or 2 is followed, embryoids will develop in groups and each embryoid within one group (derived from one pollen grain) will have the same genotype (8).

3.2. Embryoids from somatic tissue

Embryoids arising from somatic tissues also follow a variety of pathways:

1) The embryoid arises directly from an epidermal cell of the explant without an intervening callus stage. This lack of callus formation assures genetic stability, and thus true-to-type propagation. Unfortunately this type of embryoid formation generally only occurs in explants from juvenile parents. It has occured in embryo explants of *Thea* (29), *Zizyphus* (2), *Crataegus* (24), *Vitis* (1) and Populus (30).

2) The explant forms a callus first. This callus subsequently becomes embryogenic. This type of development occurred in cultures of *Actinidia* (16), *Citrus* (25), *Eriobotrya* (4) and *Prunus* (19). The embryoids develop either on or near the surface of the callus or in its interior. One has to consider the possibility that embryoids that have developed from callus may show more clonal variation than those that have developed without an intervening callus phase.

3) The embryoid arises directly from cells of the vascular tissue of the leaf. This occurred in sections of young leaves of mature *Mycetia* trees cultured on MS medium supplemented with BA 1mg/l and NAA 0.1 mg/l. After 20 - 30 days in culture embryoids started to emerge from the veins of the cultured leaves (personal communication, Lu YL). This is the ideal type of development, firstly, clonal variation is minimal because there is no callus phase, and secondly the explants can be taken from trees that have been thoroughly field tested.

4) Embryoids originate from multicellular mass, each of which has developed from a single cell. Somatic tissues of Hevea (Chen ZH, unpublished) and Eucalyptus form globular cell masses, each originating from a single cell. These masses become embryogenic (22).

4. ABNORMAL EMBRYOIDS AND MEASURES TO PREVENT THEIR FORMATION

During embryogenesis the environmental conditions are not always optimal for all of the embryoids. This results in abnormalities in these embryoids that interferes with their development into normal plantlets. The percentage of embryoids in anther cultures of the rubber tree that is abnormal is 40 - 50 %; for the other tree species the percentage is even higher. Normal embryoids have two cotyledons, a hypocotyl, a terminal shoot primordium and a root primordium. In abnormal embryoids one or more of these are either lacking or underdeveloped.

4.1. Abnormal embryoid types

4.1.1. Abnormal globular, heart, torpedo, and rod shaped embryoids. These embryoids have no terminal shoot primordium. They germinate precociously and often only form a root.

4.1.2. Embryoids with multiple meristems. These embryoids form two or more weak shoots, but only one root. Secondary embryoids that form on the primary embryoid often lack proper polarity.

4.1.3. Embryoids with abnormal root and shoot poles. When the root pole dominates, the embryoid develops a bulky root and a hypocotyl that is structured much like the root. The shoot pole remains underdeveloped, only forming a tiny green dome which sometimes turns into a callus. This type of development is common in cultures of the rubber tree. When the shoot pole dominates, the cotyledons enlarge considerably while the hypocotyl and root remain underdeveloped. Neither the root- nor the shoot-dominant embryoids develop into normal plantlets.

4.1.4. Embryoids with abnormal cotyledons. These embryoids may have only one cotyledon, or many cotyledons which appear in a rosette or are fused into a trumpet-like structure. Some of these embryoids are capable of developing into normal plantlets.

4.1.5. Vitreous embryoids. These embryoids become transparent, they fail to develop into plantlets and eventually die because of necrosis.

4.2. Measures to prevent formation of abnormal embryoids

When culture conditions are improved, the percentage of embryoids that develop into normal plantlets will increase. For example, in anther cultures of the rubber tree the frequency of plantlet formation from embryoids has been increased from 3 to 30 % by gradual improvements in the culture conditions. The following factors are important for normal development.

4.2.1. Embryoid maturation. Embryoids should not be transferred to plantlet forming medium before they have matured properly, i.e., before they have set a well developed terminal bud. Embryoids developing from rubber tree pollen form a root and cotyledons after about 30 days on differentiation medium. However, the stem and the terminal bud primordium does not develop till after 60 - 90 days in culture.

4.2.2. Hormone adjustments. Cultures should be observed carefully, and at the first sign of abnormal development corrective changes should be made to the hormone balance of the medium. Special attention should be paid to the development of the shoot pole.

4.2.3. Removal of toxins. Accumulation of toxins, in particular phenolics in the medium can cause abnormalities in embryoid development. Abnormal development can often be avoided or reduced by adding activated charcoal or polyvinylpyrrolidone (PVP) to the medium.

4.2.4. Removal of abnormal embryoids from culture. By removal of abnormal, in particular browning embryoids from the cultures at regular intervals, survival and growth of normal embryoids is improved.

REFERENCES

1. Bao XZ, Zhang JQ, Yu HH, Zou CJ, Li PF: A Preliminary Research on the Embryogenesis of Grape. Journal of Shandong University (natural sci.ed), No. 3, 1-6, 1980 (in Chinese with English abstracts).
2. Chen WL, Guo DH: Embryoid Formation of Zizyphus jujuba var. spinosus Hu in Tissue Culture. Acta Phytophysiologia Sinica Vol. 7, No. 1, 83-84, 1981 (in Chinese with English abstracts).
3. Chen ZG, Wang MQ, Liao HH: The induction of Citrus Pollen Plants in Artifical Media. Acta Genet Sin 7 (2): 189-191, 1980 (in Chinese with English abstracts).
4. Chen ZG, Lin SQ, Lin QL: A Preliminary Report on the Formaction of plantlets from Endosperm Culture of Loquat. Journal of Fujian Agricultural College, Vol. 12, No. 4, 343-346, 1983 (in Chinese with English abstracts).
5. Chen ZH, Chen FT, Qian CF, Wang CH, Zhang SJ, Xu XE, Ou XH, He YT, Lu ZM: A Process of Obtaining Pollen Plants of Hevea brasiliensis Muell.-Arg. Sci Sin 22: 81-90, 1979.
6. Chen ZH, Qian CF, Cen M, Xu XE, Xiao YL: Recent Advances in Anther Culture of Hevea brasiliensis Muell.-Arg. Theor Appl Genet 62: 103-108, 1982.
7. Chen ZH, Qian CF, Xu XE, Deng ZT: Anther Culture of Rubber Tree and Sugarcane. In: Proceedings of 5th Intl Congr Plant Tissue and Cell Culture, The Japanese Association for Plant Tissue Culture, pp 533-534, 1982.
8. Chen ZH: 13 Induction of Androgenesis in Hardwood Trees. In: Cell and Tissue Culture in Forestry, J.M. Bonga, Don J. Durzan ed. pp. 247-268, 1987.
9. Cui YF, Gong Z: Tissue Culture of Oil Palm. In: Tissue Culture of Woody Plant and its Application, Chen Zhenghua ed. High Education Press, Beijing, 466-480, 1986 (in Chinese).
10. Eriksson T: Studies on the growth requirements and growth measurements of cell culture of Haplopappus gracilis. Physiologia Pl 18 976-993, 1965.
11. Fan YH, Zang SY, Zhao JF: Induction of Haploid Plant in Lycium chinense Mill. and L. barbarum from Anther Culture. Hereditas (Beijing) 4(1): 25-26 1982 (in Chinese).

12. Fei KW, Xue GR: The Induction of Haploid Plant of Malus pumila Mill. "Yuanshuai". Acta Hortic Sin (4): 41, 1981 (in Chinese with English abstracts).
13. Fu LF, Tong DY: Induction Pollen Plants of Litchi Tree (Litchi chinensis Soon). Acta Genet Sin 10(5): 369-374, 1983 (in Chinese with English abstracts).
14. Gamborg, OL, RA Miller and K Ojima: Nutrient requirements of suspension cultures of soybean root cells. Exp. Cell Res. 50, 151-158, 1968.
15. Gu SR, Gui YL, Xu TY: Effects of Physical and Chemical Factors on Induction Frequency of the Pollen Plantlet and Changes in Starch Formation in Anthers. Acta Bot. Sinica, Vol. 26, No. 2, 156-162, 1984 (in Chinese with English abstracts).
16. Gui YL, Mu XJ, Xu TY: Studies on Morphological Differentiation of Actinidia chinensis Planch in vitro. Acta Botanica Sinica Vol. 24, No. 3, 1982 (in Chinese with English abstracts).
17. Halperin W: Embryos from somatic plant cells. In: Control Mechenisms of Cellular Phenotypes. Padykula, H.A. ed., Symposia of the International Society for Cell Biology, Academia Press, 9: 169-191, 1970.
18. Lin SK, Ji DF, Qin Z: Obtaining Haploid plantlets of Morus by Anther Culture. Sintia Sinica (B), No. 3, 280-287, 1987 (in Chinese with English abstract).
19. Liu SQ, Liu JQ: Callus Induction and Embryoid Formation in Endosperm Culture of Prunus persica. Acta Botanica Sinica, Vol. 22, No. 2, 198-199, 1980 (in Chinese with English abstracts).
20. Murashige T, and F Skoog: A revised medium for rapid growth and bioassays with tobacco tissue culture. Physol. Plant, 15: 473-497, 1962.
21. Nitsch JP and C Nitsch: Haploid plants from pollen grains. Sciences 163 (3862): 85-87, 1969.
22. Ouyang Q, Peng HZ, Li QQ: Studies on the Development of Embryoid from Eucalyptus Callus. Scientia Silvae Sinicae, Vol. 17 No. 1, 1-7, 1981 (in Chinese with English abstracts)
23. Pan JL, Wang PX, Gao RL, Fan H, Tu LZ: Tissue Culture of Elaeapnus angustifolia L. In: Tissue Culture of Woody Plant and its Application, Chen Zhenghua ed. High Education Press, 444-455, 1986 (in Chinese).
24. Wang JS, Yi G: Induced Embryoids of Young Embryo in Crataegus pinnatifida Bge. Hereditas (Beijing), 4(1): 29-30, 1982 (in Chinese).
25. Wang TY, Zhang JR: Triploid Citrus Plantlet from Endosperm Culture. Sintia Sinica, (4); 452-455, 1978 (in Chinese).
26. Wei WX, Yang YQ: The Induction of Embryoids from Longan Cotyledons and The Culture of Test-tube Plant. Journal of Fujian Teacher University, No. 2, 102-106, 1981 (in Chinese with English abstracts).
27. Yang YQ, Wei WX: Induction of Longan Haploid Plantlets from Pollens Cultured in Certain Proper Media. Acta Genet Sin 11: 288-293, 1984 (in Chinese with English abstracts).

28. Yen MQ: Successful Induction of Embryogenesis from Somatic Cells in Oil-tea. (Linye Kejitongxun), No. 12, 18, 1980 (in Chinese).
29. Yen MQ, Chen P: Studies on Development of Embryoid from The Culture Cotyledons of *Thea sinensis* L. Scientia Silvae Sinicae, Vol. 19, No. 1, 25-29 1983 (in Chinese with English abstracts).
30. Zhang LG, Zhang YH, Wang FC, Liu SL: Induction of Somatic Embryogenesis in *Populus*. Scientia Silvae Sinicae, Vol.17, No. 4, 426-427, 1981 (in Chinese with English abstracts).
31. Zou Changjie, Li Perfen: Induction of pollen plants of *Vitis vinifera* L. Acta Bot Sin 23: 79-81 (in Chinese with English abstracts).

Table 1. Explants used for induction of intact plants through in vitro embryogenesis from somatic cells

Explant	Common name	Genus	Explant	Common name	Genus
Young and mature embryos	Eucalypts	*Eucalyptus* (22)	Endosperm	Kiwi fruit	*Actinidia* (16)
	Oil palm	*Elaeis* (9)		Citrus	*Citrus* (25)
	Narrow-leaved oleaster	*Elaeagnus* (23)		Loquat	*Eriobotrya* (4)
	Hawthorn	*Crataegus* (24)		Peach	*Prunus* (19)
	Wild jujube	*Zizyphus* (2)	Sections from regenerated plants	Grape	*Vitis* (1)
	Oil-tea	*Camellia* (28)		Poplar	*Populus* (30)
	Tea	*Thea* (29)	Young leaves from mature tree		Mycetia (p*)
	Cuckoo	*Rhododendron* (p*)			
	Longan	*Euphoria* (26)			

*personal communication of Y. L. Lu

Table 2. Media for induction of somatic embryoids

Species	Step	Basic medium	Hormones (mg/l)	Organic supplement (mg/l)	Sucrose (%)	Days in culture	Final product
Zizyphus jujupa var *Spinosa* (2)	1	MS	2,4-D 2, KIN 0.5	CH 1000	3	60	Embryoids
	2	MS	IBA 0.1-0.25		3		Plantlets
Eucalyptus (22)	1	B5 or H	KIN 0.8-1, 2,4-D 2-4		4-5	20	Embryogenic calli
	2	H	BA 0.2-1, NAA 0.5		3-5	21	Embryoids
	3	Modified H	BA 0.2-1, NAA 0.5 or BA 0.2-1, IBA 0.5		3-5	9-26	Plantlets
Elaeagnus angustifolia L (23)	1	MS	ZEA 1, NAA 0.5	CH 300	2		Calli
	2	MS	ZEA 0.2, BA 2		2	Subcultured 6 times	Embryogenic calli
	3	MS	KIN 2-4, IBA 0.2		2		Embryoids
	4	MS	Shoot pretreated 2hr with 5 ppm IBA		2		Plantlets
Actinidia chinesis Planch (16)	1	MS	ZEA 3, 2,4-D 0.5-1	CH 400	2	15-30	Embryogenic calli
	2	MS	ZEA 1	CH 400	2	15	Embryoids and plantlets

Table 2 continued

<u>Crataegus pinnatifida</u> Bge (24)	1	MS	ZEA o.5, NAA 2	LH 500	5	20	Enlarged embryoids
	2	MS	BA 2, ZEA 0.5	LH 500	3		Cluster of embryoids
	3	MS	BA 0.5, KIN 0.5 IAA 0.01-0.1	LH 300	3		Green Plantlet
	4	1/2 MS	IAA 1.5		1.5		Rooting
<u>Populus</u> (30)	1	MS	2,4-D 2, KIN 1		3	30	Embryogenic calli
	2	MS or MS (macro 1/2)	BA 0.5, IAA 0.3 or BA 0.5, NAA 0.1		2-3	20-50	Embryoids, plantlets
	3	ibid	IAA 0.5		2-3	15-20	Rooting plant-lets
<u>Mycetia sinensis</u> (Hemsl.) Craib. (p*)	1	MS	BA 2, 2,4-D 1, NAA 0.2		4	30	Calli
	2	MS	BA 1, NAA 0.1		4	20-30	Embryoids and plantlets
<u>Thea sinensis</u> L. (29)	1	Modified ER	KIN 1, 2,4-D 2 or BA 0.5, 2,4-D 2	CH 300	3-5	15-30	Embryogenic calli
	2	ibid	ZEA 2, NAA 0.5	CH 300	3	30	Embryoids and plantlets

*personal communication of Y. L. Lu

MORPHOLOGICAL DEFINITION OF PHENOCRITICAL PERIOD FOR INITIATION OF HAPLOID EMBRYOGENIC TISSUE FROM EXPLANTS OF *LARIX DECIDUA*

P. von ADERKAS and J.M. BONGA, Canadian Forestry Service - Maritimes, P.O. Box 4000, Fredericton, N.B. Canada, E3B 5P7

ABSTRACT

The precise timing of explant removal has been the most important factor in the production of haploid embryoids. Over a number of growing seasons we have been able to pinpoint the best time for explant removal. This year we were able to define the best explants by date, degree-growing day, as well as in developmental terms. The condition of the cone and megagametophyte is described. Growth regulating substances used to effect maximum induction are listed. We investigated the effect of position within cones of *Larix decidua*. Megagametophytes of *Picea glauca* were also studied.

1. INTRODUCTION

Regeneration is generally more successful from female than from male gametophytic tissue of gymnosperms (2). The initiation of haploid embryogenic tissue from explants of the conifer *Larix decidua* Mill. is only possible during a very short period in the development of the female cone, i.e., between mid-June and the beginning of July (10, 13). Defining the suitable period of explant removal in terms of time is not appropriate if the intention is to predict the active period accurately, because development of cones varies in time from year to year. Cone maturation may also vary from tree to tree (7). Latitude and altitude are other factors that may delay or advance cone maturation.

The beginning of the period during which the excised megagametophyte responded best to our treatments corresponded with fertilization. However, fertilization is not a useful feature on which to base a morphological guide for explant selection, since it is difficult to detect. During 1985 and 1986 we made notes and observations on cones and megagametophytes, and consequently devised a set of morphological indicators which we tested in 1987. In addition, we extended some of the indicators to cones of white spruce (*Picea glauca* (Moench) Voss), a previously untried species.

M. R. Ahuja (ed.), Somatic Cell Genetics of Woody Plants, 29–38.

In the case of white spruce we also tried to eliminate the possibility of fertilization itself playing a role in megagametophyte inducibility, by bagging the female cones prior to pollination. We also picked three different trees of white spruce with different levels of self-fertility in an attempt to ascertain the possible role of lethal genes in megagametophyte regeneration.

Finally, positional effects of megagametophyte location were assessed within cones of *Larix decidua*, since differences of megagametophyte size had been observed in the past.

2. PROCEDURE

2.1. Materials and methods

2.1.1. *Description of source material*. Cones of *Larix decidua* were collected in 1985-1987 at the same time as explant material was removed from other cones for experiments to induce haploid embryogenic callus on a variety of media (13). The cones were all from one tree located on the campus of the University of New Brunswick, Fredericton. The three trees of white spruce were located at the Acadia Experimental Forest of the Canadian Forestry Service. These trees had been previously assessed with regard to their ability to set seed following selfing (3). Two of the trees represent the extremes, self-compatible and self-incompatible, and the third tree is genetically inbetween. The cones we used were bagged early in May to prevent pollination by white spruce. We used mentor pollen (Norway spruce) to pollinate the cones when they were ready a few weeks later. This was done because megagametophytes of spruce fail to develop beyond the early stages unless pollination has taken place. The tree is not particular about the type of pollen it receives (9).

2.1.2. *Establishment of morphological markers*. In 1985/6 fresh material of *L. decidua* was dissected and the lengths and widths of individual bracts measured. The degree of attachment of the ovuliferous scale was noted, i.e., whether it was free or not. The developing seed coat was cut open and the megagametophyte removed. After separating it from attached nucellar tissue, the megagametophyte length was measured. The megagametophyte was cut in half longitudinally (and occasionally transversely) and the accumulation of storage products was assessed using potassium iodide (IKI) for starch, and osmium tetroxide (OsO4) for lipid (4) determination. The development of various features within the megagametophyte, such as archegonia, embryos, and the corrosion cavity, a lysogenous zone into which the embryo grows as it matures, was also noted. Bracts were fixed in formalin:acetic acid:alcohol (FAA) and processed for paraffin sectioning (4). Sections were dewaxed and stained in aqueous rose bengal (5) as a general stain, or specifically for starch with IKI, or lignin with phloroglucinol/HCl (4).

2.1.3. Application of morphological markers. In 1987, we used the bract and megagametophyte measurements, as well as the condition of the megagametophyte to guide our selection of suitable explant material of L. decidua. We did not use date, as we had in the previous years. Since we had no information on appropriate collection dates for white spruce, we used morphological criteria, i.e., starch localization and corrosion cavity formation, to determine optimal timing of explant removal. Bract and megagametophyte measurements of Picea were determined in each of the collections.

2.1.4. Degree-days. Growing degree-day records (heat accumulation for days since April with temperatures above 5°C) were supplied by the local office of Environment Canada at the meteorological station at the Fredericton Airport. We compared these with the three years data on the dates of maximum induction of megagametophyte explants.

2.1.5. Megagametophyte position. Megagametophytes of larch were removed from 10 cones from collections showing the appropriate morphological markers. Five were taken from the bottom, middle, and top of each cone. Their lengths were measured and then they were plated onto medium in the usual way.

2.1.6. Statistics. Sample sizes are indicated in the appropriate captions for graphs and tables. The significance of results (p 0.01) was established by analysis of variance (ANOVA). Letter symbols are used to distinguish between the groups of data which differ at the preordained level of significance. Initiation is defined as the first macroscopic evidence of regeneration. Ability of the megagametophytes to respond to induction was assessed after three weeks.

2.1.7. Induction. In 1985 and 1986 megagametophytes (minus embryos) were placed on full strength LM medium supplemented with 1000 mg/l casein hydrolysate, 500 mg/l glutamine, 3% sucrose, and 0.1, 0.2, or 2.0 mg/l 2,4-dichlorophenoxyacetic acid (2,4-D). In 1987, all of the specimens were placed on half-strength LM medium with the same level of supplements listed above. Light and temperature conditions were as described elsewhere (13).

3. RESULTS

3.1.1. Morphological markers of Larix decidua. Certain events of female cone development were clearly not linked in any way to haploid embryogenic development of isolated explants. We knew, for example, that cones at the time of meiosis (generally in early May), or at the time of the free nuclear stage of megagametophyte development (Fig. 1A) in mid-May were not inducible (Nagmani and Bonga, unpublished). Megagametophytes developed as described by Schopf (11): the free nuclear, or coenocytic stage was followed by cellularization, initiation and maturation of the archegonia, fertilization, and onset of corrosion cavity formation. Induction in our experiments of 1985 and

1986 occurred over a 4 week period following, but not including archegonial maturation (13). This period coincided with fertilization and corrosion cavity formation. Afterwards, the inducibility of the explanted megagametophytes declined dramatically. Tissue of megagametophytes in which embryos approached maturity only gave rise to a non-embryogenic green callus, as opposed to the white embryogic tissue seen in the previous weeks and described in detail earlier (14). In terms of cone development fertilization was the period in which the bracts and megagametophytes reached maximum length (Table 1). Ovuliferous scales, which until and including archegonial formation were firmly attached to the subtending bract, became detached. Furthermore, ovule tissue destined to become the seed coat started forming lignin the following week (Fig.1B). This also corresponded to a change in starch content from high to low from one week to the next. The megagametophyte reached its maximum length after the archegonia matured. There were also other noticeable changes. The general appearance of the megagametophyte when archegonia were first distinctly visible in whole or bisected material was of a watery, translucent tissue (Fig.1C). By the following week the megagametophyte was beginning to change to opaque whitish. In bisected and sectioned megagametophytes, it was further apparent that a starch rich area had developed immediately surrounding the archegonial jackets. In some there was already an extension of this starch in a conical zone directed towards the chalazal end of the megagametophyte (Fig.1D). By the next week, this conical zone had degenerated into a cavity, often filled with a young developing embryo, and the starch was now restricted to a zone adjacent to the cavity, but not extending completely to either the lateral or chalazal edges of the megagametophyte (Fig.1E).

3.1.2. Predictive value of morphological markers for *Larix*. In 1987, we frequently sampled the same tree of *L. decidua* until it was apparent that the megagametophyte had begun to accumulate starch adjacent to and below the archegonial jacket, a stage which corresponded to transition from translucence to opaqueness of the megagametophyte. In addition, the ovule consistently showed the earliest evidence of lignification of the seed coat, and the bracts reached their maximum size. Ovuliferous scales were also readily detached by this time. These megagametophyte explants were placed on induction medium and readily formed a white embryogenic callus even though the explanted material was removed two weeks earlier than in previous years (Table 2). In 1987, we experienced an unusually warm spring and early summer. If we compare the degree-days for the three year period of these experiments, it becomes apparent that the number of degree-days required to get a responsive megagametophyte tissue is between 350 and 400, and that clearly the early spring in 1987 has temporarily advanced development of the megagametophyte tissue (Table 2 and Fig. 1F).

3.1.3. _Predictive value of morphological markers of Larix decidua on inducibility of Picea glauca._ There were evident differences between the trees in their ability to become induced. Trees 1 and 2 were responsive on the first date sampled, but tree 3 was relatively unresponsive on the first date (Table 3). Trees 1 and 2 had megagametophytes which were already well-endowed with starch by June 16, although there had been no evidence of it the week previous. Tree 3 had many clear megagametophytes while the other two had opaque megagametophytes on the day that all three trees were sampled.

As megagametophyte length increased, the response of the megagametophytes to the induction medium increased. Because mentor pollen were used to pollinate the _P. glauca_ cones, the megagametophytes are unfertilized. It is clear that such megagametophytes are easily induced. There does not appear to be any difference in induction between the trees once a suitable stage of development has been reached. Almost all megagametophytes of _P. glauca_ produced a response once they showed evidence of a corrosion cavity (Table 3). At this time other events were also noted such as the lignification of the seed coat and an increase in megagametophyte lengths. As in _Larix_, lignification occurred at the time of accumulation of starch reserves in the megagametophyte, but in _P. glauca_ this was not linked to fertilization. We never observed embryos in the corrosion cavity of white spruce. Tree 1 had a level of low self-fertility, tree 2 was intermediate and tree 3 was high (trees 553, 554 and 772 in Coles and Fowler's study (3)). There was not much of a difference between them regarding the parameters tested.

Table 3 also provides evidence that corrosion cavity development occurs independent of the act of fertilization in _P. glauca_. Indicators such as seed coat lignification and ovuliferous scale separation from the bract were equally useful in _Picea_ and _Larix_. Bract measurements of white spruce were not a good indicator, since the variation in our samples between and within trees was quite large (not shown).

3.1.4. _Megagametophyte position in Larix decidua._ In previous experiments we had only used megagametophytes from the central portion of cones, since they displayed none of the irregularities seen at either extremity. In 1987, we decided to evaluate the actual differences between the top, middle, and bottom of the cone at three dates within the phenocritical period of inducibility as defined by our markers. Megagametophyte lengths were consistently and significantly larger in the middle portion (Table 4). There was no difference between the top and the bottom. The response of the megagametophpytes was also dictated by position. Those chosen from the middle almost always

TABLE 1. Bract and megagametophyte measurements in relation to the developmental state of the megagametophyte and its inducibility (n=20)

Date	State of explant	Bract Length (mm)	Bract Width(mm)	Megagametophyte Length (mm)
June 16 1986	archegonia	8.6 ± 0.8	8.3 ± 1.4	2.16 ± .32
June 23 1986	corr. cav.	9.4 ± 0.6	10.3 ± 1.3	2.67 ± .37
June 30 1986	corr. cav.	9.6 ± 0.8	10.5 ± 1.6	2.66 ± .21

TABLE 2. Response (+ or -) of megagametophytes removed at various dates and their equivalent accumulated degree days in 1985, 1986, and 1987 when placed on induction medium

Date	Year 1985	Year 1986	Year 1987
June 8 1987			(+) 333
June 13 1985	(-) 331		
June 15 1987			(+) 397
June 16 1986		(-) 363	
June 20 1985	(+) 399		
June 22 1987			(+) 479
June 23 1986		(+) 427	

TABLE 3. Response of megagametophytes of three trees of white spruce in relation to several markers

Tree	Date	State of explant	starch (+ or -)	meg. length(mm)	response (x/40)
1	June 16 1987	corr. cav.	+	2.30 ± 0.08	35/40
	June 23 1987	corr. cav.	+	2.25 ± 0.11	39/40
2	June 16 1987	corr. cav.	+	2.17 ± 0.10	35/40
	June 23 1987	corr. cav.	+	2.35 ± 0.18	38/40
3	June 16 1987	archegonia	+	1.66 ± 0.15	10/40
	June 23 1987	corr. cav.	+	2.30 ± 0.20	36/40

TABLE 4. Megagametophyte lengths of samples removed from the top, middle and bottom portions of cones of *Larix decidua* and their rate of induction (%). (n = 40)

Location	Date June 8 1987	Date June 15 1987	Date June 22 1987
Bottom	2.20 ± .44b (9)	2.44 ± .31b (7)	2.79 ± .32b (-)
Middle	2.84 ± .17a (33)	3.23 ± .23a (33)	3.34 ± .22a (-)
Top	2.10 ± .37b (16)	2.43 ± .40b (20)	2.88 ± .30b (-)

responded, those from the top were second best, whereas those from the bottom were consistently poor.

4. DISCUSSION

Induction of megametophytes to form haploid embryogenic tissue is easily brought about if the choice of explant material is made carefully. It is generally true in conifer tissue culture that tissues collected during transition periods in which there is increased morphogenetic activity *in vivo* are given to increased plasticity in culture (1). The role of descriptive work in establishing a suitable choice of explant is consequently enhanced. In *L. decidua* the period of inducibility varied within the three years of our experiments. The variability of our single tree from one year to the next was however linked to the yearly differences in accumulation of heat, or degree-days. When we consider *P. glauca* as well, we can see that among the three trees, one tree (Tree 3) was slower than the others. This may be a genuine difference between the trees, or an effect of crown position. Measuring bract widths and lengths is only useful if the variation in cones in the tree, such as in our tree of *L. decidua*, is low. In *P. glauca* the variation was too high for this feature to be of any use. Megagametophyte length is a reasonable indicator of when to excise explants in both *P. glauca* and *L. decidua*. A threshold length of approximately 2 and 2.5 mm must be attained, respectively. The megagametophytes most suitable for induction are those which have achieved their maximum length. Megagametophytes of *L. decidua* are not equal in size within the cone. Their largest dimensions are in the middle portion. As a result, choice of megagametophytes is narrowed to a particular region on a cone at a particular time. The megagametophytes from the bottom always did poorly, even though they eventually attained lengths at which megagametophytes of the middle had already been responsive the previous week. The top megagametophytes are better than those at the bottom, even though they do not differ in size. A positional effect independent of megagametophyte length or appearance was quite noticeable. On June 8, 1987 megagametophytes at the top and bottom were clear, but those in the middle opaque as a result of accumulation of storage reserves. However, by the next week, the bottom megagametophytes had caught up in development, having both generated starch reserves and begun formation of the corrosion cavity. These bottom megagametophytes, though similar to all the middle gametophytes of the previous week, were less inducible. Our morphological criteria are therefore only applicable to the middle portion of the cone, which is also the region in which 80% of the megagametophytes are located.

In previous years (1985/86) we had to sample over a six week period to be certain of getting suitable explant material. Using the developmental markers to guide our collection, we were able to reduce this to two weeks, allowing us to collect substantially more at the phenocritical period for other experiments on initiation.

Corrosion cavity formation, a megagametophyte developmental event found in all gymnosperms, was a particularly useful marker. The usefulness of markers was particularly appreciated in 1987, since an unusually warm spring resulted in much earlier development of the megagametophyte. Without the markers, we could easily have missed the period of maximum indicibility in both of the species tested. Among the three trees of *P. glauca* there was very little difference in response to the induction process.

Megagametophytes of suitable size and appearance responded well, and no obvious differences were observed beyond those attributable to differences in maturation of megagametophytes. On the basis of many morphological studies of embryology in the Abietineae it was well-established that megagametophpyte development is a very conservative process (12). This gives us reason to expect that megagametophyte length, and starch distribution within the organ may serve as markers for other Abietinae as well. Organogenesis from regenerated tissue has also been observed in *Picea abies* (L.) Karst. megagametophytes, about one month after fertilization, a period in which Huhtinen (6) noted that the first lipid droplets and starch grains appeared. This situation is much as we described, but in gymnosperms other than Abietineae the events which allow induction may be quite different. In the cycad, *Zamia*, megagametophytes excised two months before fertilization had a high-response, but female tissue removed earlier, or later at the time of fertilization were unresponsive (8). Therefore, it would appear that selection of material must follow protocols appropriate to the species employed.

5. ACKNOWLEDGMENTS.

The technical assistance of Brenda Dow, Mike Morrisey and Anne McInnis is greatly appreciated.

6. REFERENCES

1. Bonga, J.M., and P. von Aderkas (See this volume)
2. Bonga, J.M., von Aderkas, P., and D. James (1987) Potential application of haploid cultures of tree species. *In* Genetic manipulation of woody plants, East Lansing, Michigan (in press).
3. Coles, J.F., and D.P. Fowler (1976) Inbreeding in neighboring trees in two white spruce populations. Silvae Genet. 25:29-34.
4. Gerlach, D., (1977) Botanische Mikrotechnik, 2nd ed., Georg Thieme, Stuttgart.
5. Graham, S.O. (1964) Differential staining of *Arceuthobium* tissues in the host. Phytopathology 54:1433.
6. Huhtinen, O. (1976) *In vitro* culture of haploid tissue of trees. In XVI IUFRO World congress, Division II, Forest plants and forest protection, Norwegian Forest Research Institute, N-1432 AS-NHL, Norway, pp. 28-30.

7. Kosinski, G. (1986) Megagametogenesis, fertilization and embryo development in *Larix decidua*. Can. J. For. Res. 16:1301-1309.
8. LaRue, C.D. (1954) Studies on growth and regeneration in gametophytes and sporophytes of gymnosperms. Brookhaven Symp. Biol. 6: 187-208.
9. Mikkola, L. (1969) Observations on interspecific sterility in *Picea*. Ann. Bot. Fennici 6:285-339.
10. Nagmani, R., and J.M. Bonga (1985) Embryogenesis in subcultured callus of *Larix decidua*. Can. J. For. Res. 15: 1088-1091.
11. Schopf, J.M. (1943) The embryology of *Larix*. Ill. Biol. Monogr. 19: 1-97.
12. Singh, H. (1978) Embryology of gymnosperms. Gebrüder Borntraeger, Berlin, Stuttgart.
13. von Aderkas, P., J.M. Bonga, and R. Nagmani (1987) Promotion of embryogenesis in cultured megagametophytes of *Larix decidua*. Can. J. For. Res. (in press).
14. von Aderkas, P., and J.M. Bonga (1987) Formation of haploid embryoids of *Larix decidua*: early embryogenesis. Amer. J. Bot. (in press).

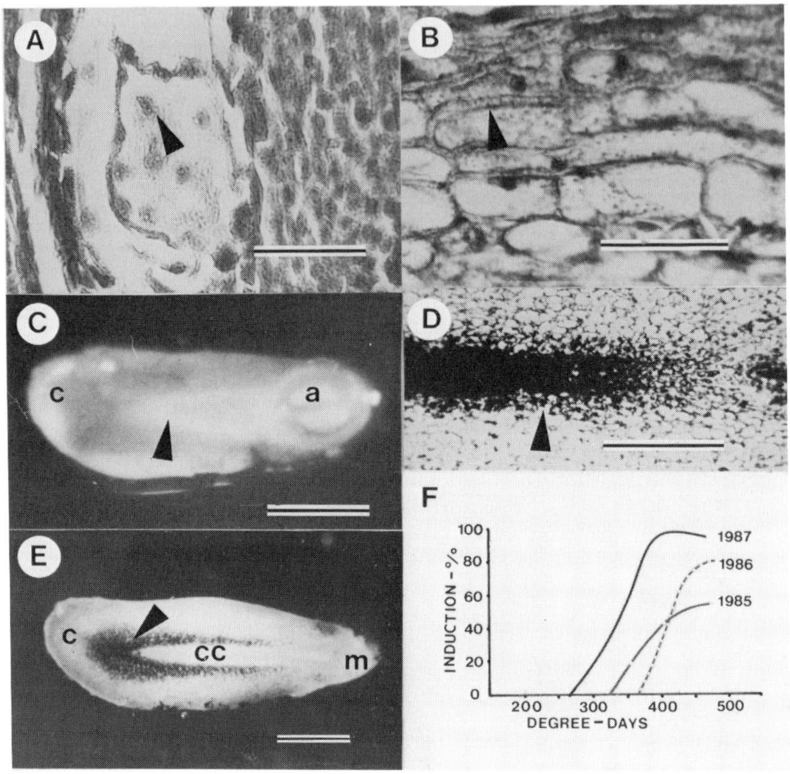

Figure 1. Larix decidua A) Section of ovuliferous scale from mid-May showing the early free-nuclear stage of the megagametophyte. Stained with rose bengal. Bar = 100 um. B) Section of the future seed coat stained with phloroglucinol/HCl. The inner cells (arrows) are thickened substantially and stain positive. The cells are no longer cytoplasmically dense. Bar = 100 um. C) Hand-section of a fixed megagametophyte of June 16, 1986 at the time of archegonial maturation, but before corrosion cavity formation. Note the absence of storage products in the highly vacuolate cells of the megagametophyte surrounding the site of corrosion cavity formation (arrow). Stained with rose bengal. Bar = 0.5 mm. D) Longitudinally bisected megagametophyte at the time of corrosion cavity formation. Staining with IKI for starch reveals the corrosion cavity (arrow) in the centre. Bar = 250 um. E) Longitudinally bisected unstained megagametophyte at the time of fertilization (June 8, 1987). Note the corrosion cavity (cc) zone extending from the micropylar (m) towards the chalazal (c) end. Also visible is a zone surrounding the corrosion cavity which stains positively with IKI for stand (arrow) Bar = 0.5 mm. F) Induction percentage in relation to degree-days in 1985, 1986, and 1987.

PRODUCTION OF HAPLOID PLANTLETS IN CULTURES OF UNPOLINATED OVULES OF HEVEA BRASILIENSIS MUELL.-ARG

CHEN ZHENGHUA, LI WENBIN, ZHANG LIHUA

Institute of Genetics, Chinese Academy of Science, Beijing, China

XU XUEN, ZHANG SHIJIE

Baoting Institute of Tropical Crops, Guangdong, China

0. ABSTRACT

Plantlets were obtained by culture of unpolinated ovules of rubber tree, *Hevea brasiliensis*. The ovules were cultured on a dedifferentiation medium, and later transfered to a differentiation medium. Cytological observations indicated that the plantlets tested were mixaploids.

1. INTRODUCTION

Haploid plants of *Hevea brasiliensis* have since 1977 been obtained from cultured anthers (4). However, anther culture cannot be used for varieties in which the staminate flowers degenerate at an early stage. Such early degeneration occurs in several of cold resistant forms of the species. For such trees, haploid plants can only be obtained by culture of unpolinated ovules.

Haploid plants have been produced in cultures of unpolinated ovules of a number of herbaceous crops (10, 11). With tree species, haploid plant production from unpolinated ovules *in vitro* has so far been limited to poplar (9). One report from China indicates that one haploid plantlet was obtained from cultured unpolinated ovules of the rubber tree (8). However, the ploidy of this specimen was not confirmed. Below we will describe the formation of embryoids and plantlets from the haploid embryo sac cells of unpolinated ovules of rubber tree.

2. MATERIALS AND METHODS

2.1. Explant

We started culture of unpolinated ovules of *Hevea brasiliensis* in 1982. To obtain the explants, pistillate flowers were removed from inflorescences young enough to have unopened flowers. The largest staminate flowers at that time were about 2.0 - 2.5 mm long, with anthers that contained pollen in the tetrad stage. Pistilate flowers were also taken from inflorescences with staminate flowers up to 3.0 - 3.5 mm long with anthers harboring pollen mostly at the uninucleate stage and some at the binucleate stage.

2.2. Surface sterlization

Pistillate flowers were wrapped in cheescloth gauze, with gauze package containing flowers of one and the same developmental stage, that is, one treatment. Each package was soaked in a 0.1 percent mercuric chloride solution for 10 minutes and then washed 4 - 5 times in sterile water.

M. R. Ahuja (ed.), Somatic Cell Genetics of Woody Plants, 39–44.

2.3. Excision and inoculation

The ovule was removed from the young ovary. During this procedure every effort was made to avoid damage to the ovule. There are three ovules in each ovary. When small culture tubes (12 x 120 mm) were used, five ovules were placed in each of these; when large tubes (20 x 200 mm) were used, ten ovules were transfered to each tube.

2.4. Culture media

When the ovary wall is removed only little somatic tissue remains around the haploid cells of the embryo sac. Such small cells masses are generally much more sensitive to nutrient media variation than larger ones and thus need careful media adjustments.

2.4.1. Dedifferentiation medium. To obtain dedifferentiation we designed the RT (rubber tree) medium (6), and the MB medium (1, 4), which contained (in mg/l) the following:

-Macroelements: KNO3 950, NH4NO3 1650, KH2PO4 510, MgSO4.7H2O 370, CaCl2.H2O 340

-Microelements: MnSO4.4H2O 10, ZnSO4.7H2O 8.6, H3BO3 12, CuSO4.5H2O 0.025, Na2MoO4.H2O 0.25, NaEDTA 37.2 and FeSO4.7H2O 27.8

-Organic supplements: thiamine-HCl 0.5, pyridoxin-HCl 0.5, glycine 2, nicotinic acid 0.5, folic acid 1.0, coconut water 5 percent v/v, sucrose 7 percent, and agar 0.6 percent. The pH was adjusted to 5.8 before autoclaving. The hormonal supplements are mentioned later in the text.

2.4.2. Differentiation medium. Two (RT and MB) differentiation media were used. The first one contained 4/5 strength of Murashige Skoog (MS) macrosalts, double strength MS microelements, and single strength MS organic supplements. The second medium contained single strength RT medium macro elements, double strength MS microelements, and single strength MS organics. Both media contained sucrose 8 percent and agar 0.6 percent.

2.4.3. Plant forming medium. This medium contained the macroelements of MS at 4/5 strength, the microelements of MS at double strength, and the organics of MS at normal strength. In addition, the medium contained gibberellic acid (GA) 2.0 - 4.0 mg/l, indoleacetic acid (IAA), 1.0 -2.0 mg/l, 5-bromouracil 0.5 -1.0 mg/l (added to prevent callusing of the embryoids, see (6), sucrose 5 percent, and agar 0.8 percent.

2.4.4. Environmental factors. Diffuse light (12 hours/day) appeared to be adequate for callus and embryoid formation. For plantlet formation the cool white light at 2500 - 3000 lux was appropriate. All of the explants were maintained in a room with 25 - 28 C temperature.

2.5. Cytological procedures

Young leaves (0.5 - 1.0 cm long) were removed from plantlets produced in vitro between 11:30 and 12:00 in the morning, that is, at a time during which the mitotic rates in the leaves are high. The leaves were submerged in a 0.003 M solution of 8-hydroxyquinoline at room temperature for 4 hours, and subsequently stained and squashed for chromosome counting as described elsewhere (2).

3. RESULTS

The haploid cells of the embryo sac formed callus or embryoids when cultured on RT medium supplemented with either kinetin (KIN) 1.0 -1.5 mg/l or N6-benzyladenine (BA) 0.5 - 1.0 mg/l, and with a combination of naphthaleneacetic acid (NAA) 1.0 - 2.0 mg/l and 2,4-dichlorophenoxyacetic acid (2,4-D) 1.0 mg/l. Many ovules formed callus on these media within 10 days. The ovules taken from inflorescences in which the anthers in the staminate flowers contained uninucleate stage pollen produced in most callus. For example, 25 of 73 (34 percent) of such ovules inoculated during a single day in 1982 formed callus, while only a few ovules from younger or older inflorescence did so. Most calli were yellow in color and had a soft texture. After 40-50 days of culture on a dedifferentiation medium, the calli were transfered to differentiation medium supplemented with either KIN 0.5 - 1.0 mg/l or BA 0.3 - 0.5 mg/l, and with NAA 0.2 -0.3 mg/l, GA 0.5 - 0.8 mg/l.

The calli produced 24 embryoids while on the differentiation medium. When the concentration of cytokinin was at the upper limit in both the dedifferentiation and differentiation media, the calli grew rapidly but did not form embryoids. With a low level of cytokinin in the dedifferentiation medium embryoids sometimes formed directly on the ovules, that is, they did not develop from callus. However, these embryoids became abnormal after transfer to the differentiation medium, and only a few reached the torpedo stage. Embryoids which developed from callus grew in a more normal fashion. Some assumed a trumpet shape because of fusing of cotyledons, while others formed normal cotyledons. Both types had the capacity to develop into normal plantlets. In 1982, because of lack of experience, all embryoids were transfered to the plant forming medium after 60 days of culture on the differentiation medium, which, as was found out later, was too short a period to ensure normal embryoid development. These embryoids formed green cotyledons and roots, but failed to develope a normal stem.

In 1984 the experimental conditions were the same as in 1982, except that the embryoids remained on the differentiation medium for 90 instead of 60 days. Of the 77 ovules in one experiment, 27 showed a response. Of these 15 formed, without prior callus formation, one embryoid each. These embryoids grew slowly and senesced before reaching maturity. The other 12 ovules develope callus most of which was yellow and the rest green. Of the yellow calli, six formed 34 embryoids. Fourteen of these embryoids developed into trumpet type or normal cotyledonary embryoids, which matured after 90 days on the differentiation medium (Figure 1). Four of these formed normal plantlets (Figure 2) after transfer to the plant forming medium. In the next year an additional two normal plantlets were obtained by the same method.

The rubber tree is an amphidiploid (2n = 36) with a basic number of 9 and a gametic number of 18. Karyotyping was carried out with young leaves from two of the plantlets. Of the 98 metaphase plated studied, only a few had 9 (Figure 3) chromosomes; most had either 18 (Figure 4) or 27 chromosomes; a few were aneuploid and a few had 36 chromosomes (Table 1). Multipolar mitosis have been found in cytological observations, that is, there were two poles on one side of the equatorial plate, one with 9 chromosomes and the other with 18 chromosomes, and on the other side there was one pole with 9 chromosomes. In such cases cells with 27 and 9 chromosomes might be formed. Thus the plants were mixaploids, a condition which probably had arisen mosly by gradual, stepwise (by the basic number of 9) increase in the chromosome numbers during development of the embryoid and plantlet. A similar ploidy pattern was observed in plantlets derived from pollen of the rubber tree (2).

Table 1. Chromosome number in cells of young leaves of plantlets

Plant	Total number of metaphases	Chromosome number					
		9	18	19-26	27	28-35	36
1	37	3	15	1	11	3	4
2	61	0	23	6	24	5	3
Total	98	3	38	7	35	8	7

4. DISCUSSION

Ovules from inflorescences, with staminate flowers with most of the pollen in the mononucleate stage and the rest having just entered the binucleate stage, were best for the production of haploid plsntlets from the embryo sac cells. To determine the development stage of ovules directly is difficult. However, pollen development correlated well with ovule development and provided an easy and reliable marker to indicate the right time of ovule collection. A similar method was used with rice to obtain ovules at the proper stage of development for culture. Parafin sections of flowers of this species showed that the optimal stage on inoculation ranged from uninucleate to four-nucleate embryo sacs, corresponding to late uninucleate to early-binucleate pollen stage (11). Similarly, it was found that poplar ovules were at a responsive stage of development when the pollen of the staminate flowers in the same inflorescence were uninucleate (9).

Ovules of the rubber tree formed embryoids either directly, without prior callus formation, or from callus. Similarly, unfertilized ovules of poplar *in vitro* occasionally formed embryoids either directly or, more frequently, after first forming callus. Most of the poplar embryoids had poorly developed roots and did not develop into normal plantlets until after proper lateral root induction (9).

Studies of embryoid and plantlet production from pollen have shown that there are two critical phases during embryoid development. The first phase is pre-primordial, during the second one the primordia are initiated. Usually the root primordium forms earlier than the shoot primordium (5). These two stages move more slowly in some woody plants than in herbaceous plants (3, 5, 7). Embryoids developing from the haploid embryo sac cells follow a growth pattern similar to that of embryoids arising from pollen. They mature even slower than the embryoids of pollen origin. In our experiments proper plantlets formed only in embryoids left on the differentiation medium for 90 days.

REFERENCES

1. Bourgin JP and JP Nitsch: Obtention de Nicotiana haploides a partir d'etamines cultivees in vitro. Ann. Physiol Veg. 9: 337-382, 1967.
2. Cen M, ZH Chen, CF Qian, CH Wang, YT He and YL Xiao: Investigation of ploidy in the process of anther culture of Hevea brasiliensis. Acta Genet Sin 8(2): 169-174, 1981 (in Chinese with English abstract).
3. Chen ZG, MQ Wang and HH Liao: The induction of Citrus pollen plants in artificial media. Acta Genet Sin 7(2): 189-191, 1980 (in Chinese with Engrish abstract).
4. Chen ZH, FT Chen, CF Qian, CH Wang, SJ Zhang, XE Xu, XH Ou, YT He and ZM Lu: A process of obtaining pollen plants of Hevea brasiliensis Muell.-Arg. Sci Sin 22: 81-90, 1979.
5. Chen ZH: Induction of androgenesis in woody plants. In: Haploids of Highr Plants In Vitro, Hu Han and Yang Hongyuan (Eds.), China Academic Publishers, Beijing and Springer Verlag, Berlin, Heidelberg, pp. 42-66, 1986.
6. Chen ZH: Induction of androgenesis in hardwood trees. In: Cell and Tissue Culture in Forestry, Vol 2, JM Bonga and DJ Durzan (Eds.), Martinus Nijhoff Publishers, Dordrecht, pp. 247-268, 1987.
7. Fu LF and DY Tong: Induction of pollen plants of Litchi tree (Litchi chinensis Soon). Acta Genet Sin 10(5): 369-374, 1983 (in Chinese with English abstract).
8. Guo GF, XJ Jia and LX Chen: Induction of plantlet from ovule in vitro of Hevea brasiliensis. Hereditas 4(1): 27-28, 1982 (in Chinese).
9. Wu KX: Unpolinated ovary culture of poplar in vitro. In: Tissue Culture of Woody Plants and its Application, ZH Chen (Ed.), High Education Press of China, pp. 179-186, 1986.
10. Yang HY, C Zou, DT Cia, H Yan, Y Wu and XM Chen: In vito culture of unfertilized ovules of Helianthus annuus L. In: Haploids of Higher Plants In Vitro, Hu Han and Yang Hongyuan (Eds.), China Academic Publishers, Beijing, and Springer Verlag, Berlin, Heidelberg, pp. 182-191, 1986.

11. Zhou C, HY Yang, HA Tian, ZL Liu and HH Yan: In vitro culture of unpolinated ovaries in Oryza sativa. In: Haploids of Higher Plants In Vitro, Hu Han and Yang Hongyuan (Eds.), pp. 165-181, 1986.

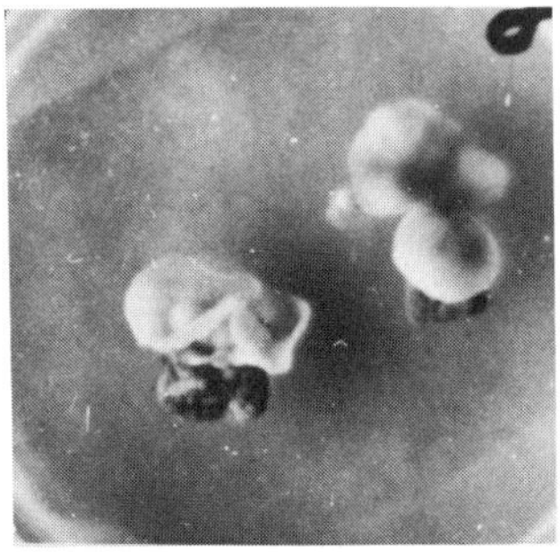

Figure 1. Embryoids from calli of unpolinated ovules of *Hevea*.

Figure 2. A plantlet from an unpolinated ovule of *Hevea*.

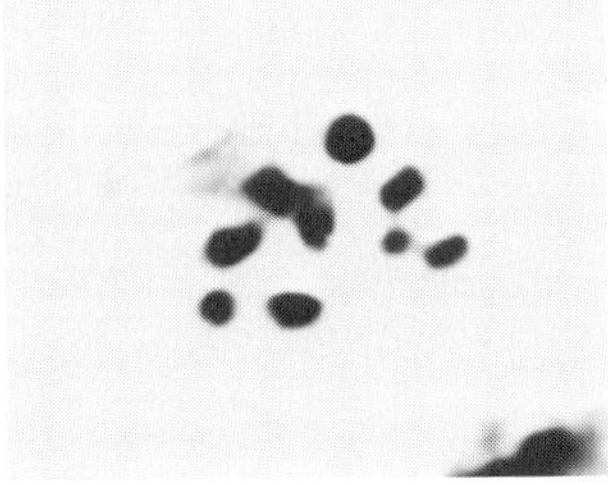

Figure 3. A cell with 9 chromosomes from a young leaf of a plantlet.

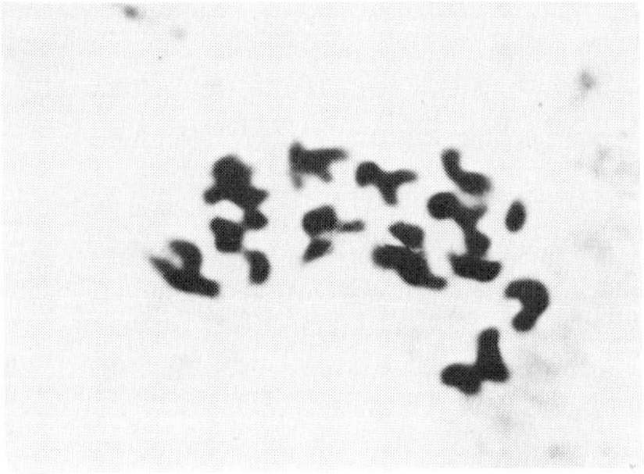

Figure 4. A cell with 18 chromosomes from a young leaf of a plantlet.

Somatic embryogenesis in tissue cultures of walnut (*Juglans nigra*, *J. major* and hybrids *J. nigra* * *J. regia*).

D. CORNU: Station d'Amelioration des Arbres Forestiers
I.N.R.A.-C.R. d'Orléans, Ardon 45160 Olivet
FRANCE

ABSTRACT

From cotyledons of immature nuts of *Juglans nigra*, *J. major* and interspecific hybrids, it is possible to obtain somatic embryogenesis. Some embryos could give whole plants, but most of them show abnormal development (swelling, chlorophyll deficiency...).

1. INTRODUCTION

In France, walnut trees are very interesting for their high quality wood production particularly for veneer and for furniture. Traditionaly we used persian walnut (*Juglans regia*) at the end of their fruit production. Now, the new orchard management technics oblige forester to produce trees only for wood production. Interspecific hybrids between black and persian walnut (*J. nigra* * *J. regia*) and other hybrids are very interesting for their soil adaptability and their fast growth (I.D.F. 1983). But today the production of hybrid nuts is very low and variable from year to year.

Vegetative propagation through cuttings is difficult (Cornu 1977). If it is possible to use micropropagation (McGranahan et al. 1987), many problems, like aging and rejuvenation (Jay-Allemand et al. 1987), contaminants, rate and quality of rooting, soil transfer..., remain important to find a technic for mass propagation.

Somatic embryogenesis could allow to avoid these problems because it is possible by this way to produce directly juvenile whole plant with apical bud and main root tip. Tulecke and McGranahan (1985) obtained somatic embryos from *J. regia*, *J. hindsii* and *Pterocarya* and Merckle et al. (1987) from *Carya illinoensis*, all species of the Juglandaceae family. The aim of this work is to show same results on other species and hybrids.

2. MATERIAL AND METHODS

Nuts were collected on mature trees (2 to 10 nuts per tree) from early August through September, they were all open pollinated. We have used 15 *J. nigra*, 3 *J. major*, 5 *J. regia* and 3 hybrids *J. nigra* * *J. regia* which produce fruits. Before they were used the nuts were stored less than 5 days in cold room (+4°C).

M. R. Ahuja (ed.), Somatic Cell Genetics of Woody Plants, 45–49.

For all experiments we used media defined by Tulecke and McGranahan (1985) for conditioning and for proliferating stages. Fruits were sterilized by flaming, cut with scalpel (or cracked when lignified), and whole embryo with cotyledons (when less than 5mm) (Fig. 1) or pieces of cotyledons were excised and placed on agar medium.

Cotyledon tissue and somatic embryos were grown in dark at a temperature of 25°C. Greening and development of somatic embryos were obtained under a photoperiod of 16 h.

3. RESULTS AND DISCUSSION

For nuts collected in September (one *J. nigra*) or in August (all *J. regia*), for which the cotyledons were fully developed, we observed no reaction. Rapidly the explants turn black, probably by oxydative process of lipids. There were also no reaction for nuts collected too early (6 *J. nigra*), for which the embryos are not enough developed (less than 1mm).

For the others, embryos were immature and ranged in development from 2mm up to 1/3 of their full size. After several weeks they reacted and produced a compact white callus. This kind of callus is similar in aspect to normal cotyledon and grows in nodular or leaf-like structures (Fig. 2).

After two tranfers on medium without growth regulator, only sixexperimented trees (2 *J. major*, 3 *J. nigra* and one hybrid) showed organogenesis or formation of embryo-like structures (Fig. 2,3 and 4). The cultures were maintained by subculturing cotyledon-like structures on this medium every 2 to 3 weeks. They are still producing somatic embryos after 8 months. Somatic embryos appear directly on these structures (Fig. 3) or on embryos which were formed previously (Fig. 2).

The embryos could be isolated and subcultured either in dark or in light. But few of them give a normal development to whole plant. Even on medium without growth regulator they frequently show swelling or callusing and continue to produce new embryos. We can obtain by this way cluster of embryos at different stages of development (Fig. 5 and 6).

In the light the culture turn green, but this process is very heterogeneous and we obtain a mosaic of green, red, white, brown,... nodular parts. For isolated somatic embryos, the reaction is more homogeneous but many of them did not turn green. They show lack of chlorophyll formation in the shoot and after 3 to 4 weeks they turn brown and die.

Only few somatic embryos complete their growth and give whole plants (Fig. 7 and 8), but they are well constituted with a main root tip and an apical bud and show a normal growth.

These results on the choice of explants to start cultures, on their development and on the formation and the subsequent growth of somatic embryos, are very similar to those obtained on other species of Juglandaceae. We show that somatic embryogenesis can be obtained also with *J. nigra*, *J. major* and above all with interspecific hybrids which are of special interest for reforestation.

REFERENCES

Cornu D.: La multiplication végétative du noyer hybride. Résultats d'une première campagne, perspectives d'avenir. Rev. for. fr 29(6), 457-463, 1977.

I.D.F.: Les noyers à bois. Institut pour le développement forestier, Paris (Ed). 132 pp, 1983.

Jay-Allemand C, Cornu D. and Macheix J.J.: Biochemical attributes associated with rejuvenation of walnut. Plant Physiol. and Biochem., (in press) 1987.

McGranahan G.H., Driver J.A. and Tulecke W.: Tissue culture of *Juglans*. in Cell and Tissue Culture in Forestry, vol 3, Bonga J.M. and Durzan D.J. (Ed). 261-271, 1987.

Merckle S.A., Wetzstein H.Y. and Sommer H.E.: Somatic embryogenesis in tissue culture of pecan. HortScience 22(1), 128-130, 1987.

Tulecke W. and McGranahan G.H.: Somatic embryogenesis and plant regeneration from cotyledons of walnut, *Juglans regia* L. Plant Sci. 40, 57-63, 1985.

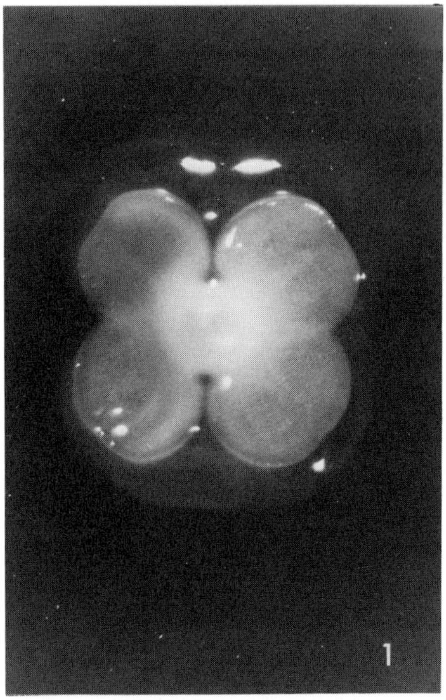

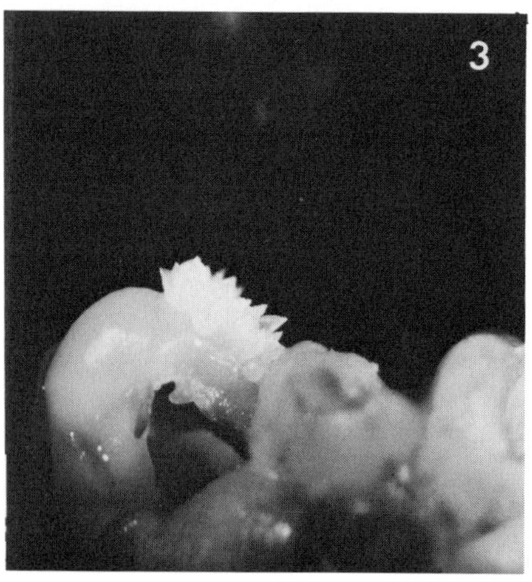

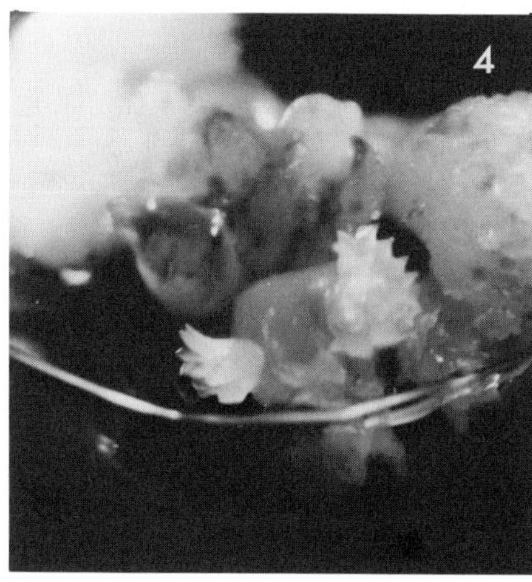

Fig. 1: Immature embryo of *Juglans nigra* on conditioning medium (size: 3mm in diameter).

Fig. 2: Cotyledon-like (top) and embryo-like (bottom) structures initiated on a swelled somatic embryo.

Fig. 3 and 4: Organogenesis on cotyledon-like structures (secondary cotyledon or somatic embryo).

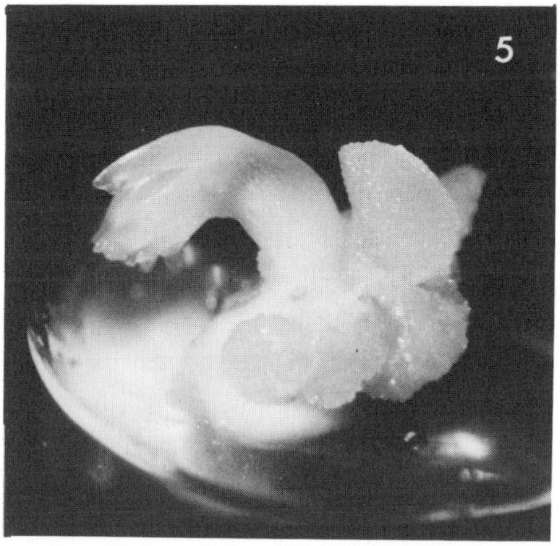

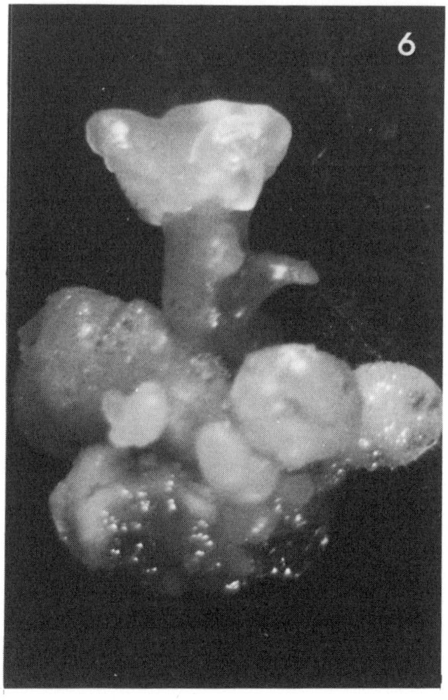

Fig. 5: Somatic embryo initiated on a globular structure (note the callusing of the other formations).

Fig. 6: Cluster of somatic embryos at different stages of development.

Fig. 7: Two somatic embryos formed on a cotyledon-like structure.

Fig. 8: The same after transfer in light (note the development of cotyledons and the vigorous root growth).

IN VITRO EMBRYOGENIC CALLUS FORMATION IN *Chimonanthus*

Lj.Radojević[1], N.Djordjević[1], M.Guć-Šćekić[2]. [1]Inst.Biol.Res. "S.Stanković", [2]Inst.Mother and Child Health, 11 000 Belgrade, Yugoslavia

Capability of uninuclear microspores of the woody sp. *Ch.praecox* L. (2n=22) for androgenesis and/or callus production in the culture of anthers from the buds in successive developmental stages (1-4) and cultured in MS (Murashige and Skoog, 1962) medium with 2,4-D, Kin., and vitamines (Radojević, 1980) was studied.

In this work, the influence of hormones (2,4-D, GA_3), sucrose, ficoll, proline and oxygenation (20, 30 and 60 min; 2^3ATA) on the formation of androgenic embryogenic callus in *Chimonanthus* anther culture, or androgenic embryos induction was examined.

In MS liquid or solid media 4 types of calli were obtained: white compact embryogenic and friable non-embryogenic white, light yellow and dark yellow. Embryogenic callus was induced on solid medium with 20-30 kg m^{-3} sucrose, 2,4-D or GA_3, Kin. (1x 10-3 kg m^{-3}, each) with or without proline (0.25 kg m^{-3}). Callus in MS medium with GA_3 was 2-3 fold more voluminous than on medium with 2,4-D, Histological sections of the anthers grown in the presence of 2,4-D revealed 2-8-celled proembryos, while multicellular embryos with distinct bipolar axes were seen in the presence of GA_3. Their further development was ceased.

Increased sucrose (60 kg m^{-3}) gave 50% anthers producing androgenic callus. Further increase to 120 kg m^{-3} completely inhibited callus production. If proline was omitted and cultures subjected to oxygenation (60 min, 2ATA), 10% anthers produced embryogenic callus. Under the same conditions of oxygenation and increased 2,4-D ($3x10^{-3}$ kg m^{-3}), 37% anthers gave embryogenic callus. Ficoll (100 kg m^{-3}) in the culture medium completely albolished embryogenic callus induction. In liquid MS medium with $3x10^{-3}$ kg m^{-3} 2,4-D and $1x10^{-3}$ kg m^{-3} Kin., certain inhibition of embryogenic callus formation was observed depending on the oxygenation period. Proline in the latter medium and oxygenation (30 min., 2 ATA) led to the inhibition of embryogenic callus. Androgenic anther response was not obtained in liquid MS medium with GA_3 and Kin. ($1x10^{-3}$kg m^{-3}) with or without ficoll. If anthers on the same medium were oxygenated (60 min, 2 ATA) slight stimulation of androgenic callus formation was observed (14% anthers).

After 6 months of embryogenic callus culture on MS agar medium with 2,4-D and Kin. ($1x10^{-3}$ kg m^{-3}), cytogenetic analyses of proembryos (Fig.1) showed haploid, diploid and tetraploid chromosomes number. After 2 yrs of this callus culture in the same medium polyploid-hypotetraploid (4n-4, 4n-6, 4n-8)chromosome numbers in 30 analysed cells were observed (Fig.2). Haploidy or diploidy were not seen. Chromosome number ranged from 30-40. Depending on the centromere position, metaphase chromosomes were separated into median, submedian and subterminal (Fig.3, A, B and C, respectively).

The similarity of cytogenetic results in long-term cultures of *Chimonanthus* and those of long-term cultures of other plant species (Krikorian *et al.*, 1983) should be emphasised. Cytogenetical anlyses of callus strains cultured in liquid MS media

M R. Ahuja (ed.), *Somatic Cell Genetics of Woody Plants*, 51–52.

are in progress.

REFERENCES

Krikorian AD, O'Connor SA and Fitter MS: Chromosome number variation and karyotype stability in cultures and culture-derived plants. In: Handbook of Plant Cell Culture. Evans DA *et al.* (Eds.) Macmilaln Publish.Co., New York, Coller Macmillan Publich. London, pp. 542-591, 1983.

Murashige T and Skoog F: A revised medium for rapid growth and bioassays with tobacco tissue cultures. Physiol.Plant. 15, 473-497, 1962.

Radojević Lj: Embryogenèse somatique et androgenèse chez certaines espèces ligneuses. Bull.Soc.Bot. France, 127, Actuel Bot. 3, 99-107, 1980.

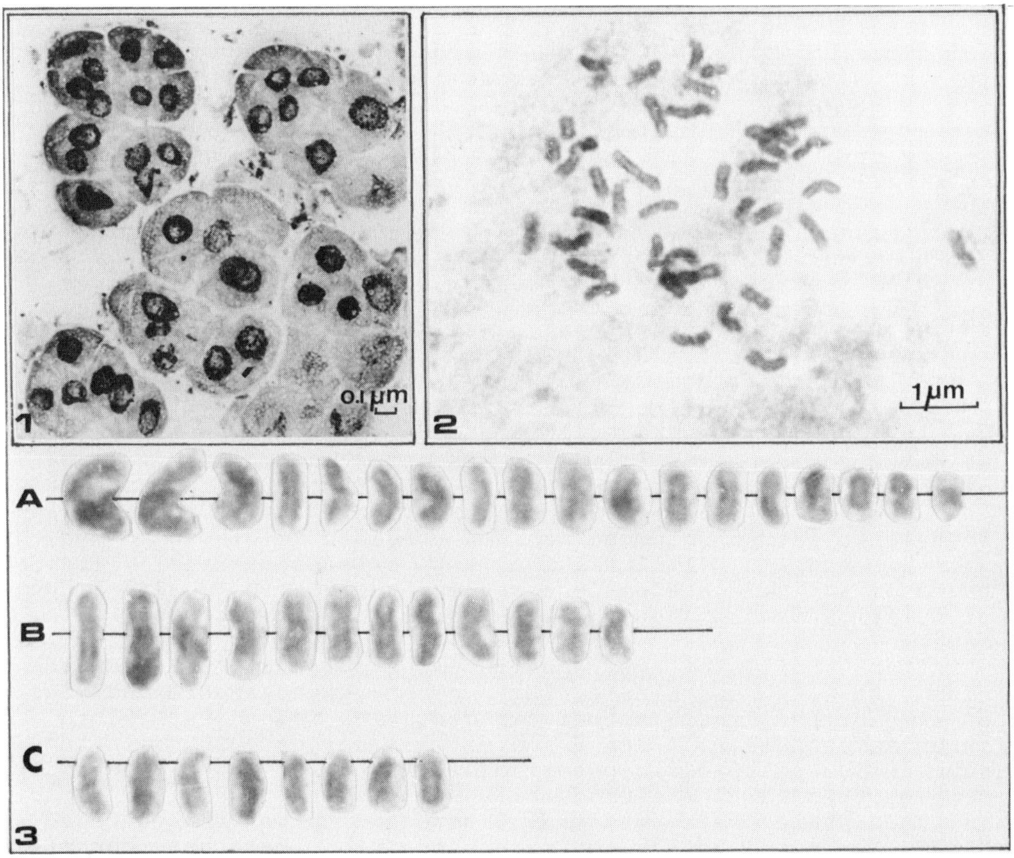

Fig.1. Proembryos in embryogenic *Chimonanthus* callus on MS agar medium, 2,4-D and Kin. $(1 \times 10^{-3} kg\ m^{-3})$.

Fig.2. Metaphase chromosomes from a proembryo culture on the same medium.

Fig.3. Karyotype showing medium, submedium, and subterminal chromosomes.

PLANT REGENERATION OF HORSE CHESTNUT BY IN VITRO CULTURE

Ljiljana Radojević, "S. Stanković" Institute for Biological Research, 11 000 Belgrade, Yugoslavia

Horse chestnut (*Aesculus hippocastanum* L.) explants were propagated *in vitro* by: 1. androgenesis (Radojević, 1978);2. meristem culture from androgenic embryos (Radojević *et al*., in press) and 3. somatic embryogenesis.

Immature embryos cultured on solid MS (Murashige and Skoog, 1962) medium containing sucrose (50 kg m^{-3}), B vitamins, 2,4-D, Kin. (3x10^{-3} kg m^{-3}; 1x10^{-3} kg m^{-3}, respectively), casein hydrolysate (0.25 kg m^{-3}), with or without proline (0.25 kg m^{-3})were used.

Two types of calli: whitish compact non-embryogenic and globular embryogenic were formed. They were separated and subcultured for further embryogenic callus growth in the same medium with reduced sucrose (30 kg m^{-3}), 2,4-D, Kin. (1x10^{-3} kg m^{-3}, each) and proline (0.25 kg m^{-3}) resulting in an increase of both callus mass and number of embryos (70-100 embryos/culture; Figs. 1-2.). Embryo development was asynchronous and globular, heart-shaped, torpedo and embryos with cotyledons were formed in the same culture. Embryos in cotyledonary stage of development were further cultured on different media for germination (Fig.3.).

Germination and development of somatic embryos into plantlets was achieved by short-term cultures on MS solid or liquid medium. On MS medium with IBA and GA_3 (1x10^{-3} kg m^{-3}) cultures kept at 4°C during 3 weeks gave 5% of germinating embryos. Proline (0.25 kg m^{-3}) added to the same medium significantly diminished number of germinating embryos to 1% (Fig.5, A and A_1). Embryos cultured in the same media at 25°C did not germinate (Fig.4, A and A_1). Cultures in MS liquid medium with IAA, GA_3 (1x10^{-3}kg m^{-3}); Glu (0.4 kg m^{-3}) and activated charcoal (AC; 0.1 kg m^{-3}) gave 57% of the germinating embryos (Fig.6.), while 65% of the embryos germinated in culture medium devoid of AC (Fig.7.).

Addition of cytokinins (BAP, Kin., zeatin, 3x10^{-3} kg m^{-3}each) inhibited plantlets formation.

Regenerated plantlets had diploid chromosome number (2x=2n=40).

Propagation of horse chestnut *via* somatic embryogenesis, androgenesis or meristem culture, represents a novel approach for the propagation of this woody species on large scale.

REFERENCES

Radojević Lj: *In vitro* induction of androgenic plantlets in *Aesculus hippocastanum* L., Protoplasma, 96, 369-374, 1978.

Radojević Lj, Druart P and Boxus P: Vegetative propagation of androgenous embryos of horse chestnut by meristem culture *in vitro*. Acta Horticulture (in press).

Murashige T and Skoog F: A revised medium for rapid growth and bioassays with tobacco tissue cultures. Physiol.Plant. 15, 473-497, 1962.

M. R. Ahuja (ed.), Somatic Cell Genetics of Woody Plants, 53–54.

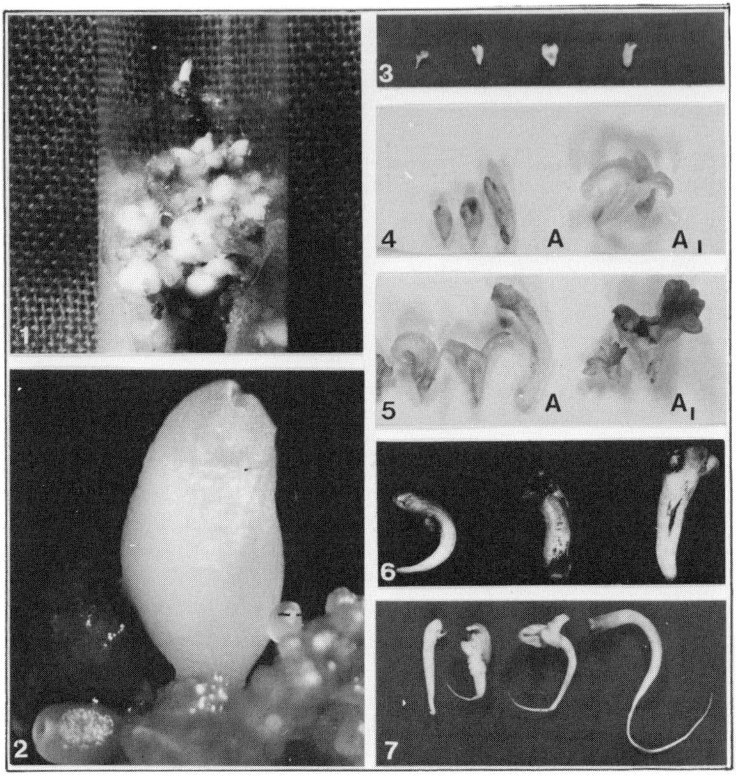

Figs.1-2. Culture of embryogenic Aesculus callus with somatic embryos in different stages of development.

Figs.3-7. Further development of somatic embryos into horse chestnut plantlets.

A MODEL SYSTEM FOR GENE TRANSFER IN CONIFERS: EUROPEAN LARCH AND AGROBACTERIUM

D.F. Karnosky and A.M. Diner
School of Forestry and Wood Products
Michigan Technological University
Houghton, MI 49931

W.M. Barnes
Dept. Biological Chemistry
Washington University
School of Medicine
660 South Euclid
St. Louis, MO 63110

1. ABSTRACT

Newly developing forest biotechnologies offer numerous opportunities for rapid and efficient gene transfer. However, conifers remain generally recalcitrant to these techniques, primarily because systems have not yet been developed for recovering whole transformed plants. There are four main components of an effective *Agrobacterium*-mediated gene transfer system. First, desirable genes and their promoters must be incorporated within the borders of the plasmid transfer DNA. Second, a suitable *Agrobacterium* vector system must be established for delivery of the plasmid into the host. Third, suitable marker genes must be available to allow for *in vitro* selection of transformed cells, tissues, or organs. Fourth, techniques for regeneration of whole plants are necessary. We have made substantial progress towards these requirements with *Agrobacterium*-mediated transfer for European larch (*Larix decidua*). We have shown several strains of *A. tumefaciens* and *rhizogenes* to be virulent to young *Larix in vitro*. Host tissues were transformed by a binary plasmid vector bearing genetic loci for synthesis of neomycin phosphotransferase (NPT) which conveys resistance to the antibiotic kanamycin. Calli generated either from hairy roots (*A. rhizogenes* pARC 8) or hypertrophic inoculated hypocotyls (*A. tumefaciens* pWB101) survived and grew on media supplemented with kanamycin to levels of 30 and 300 mg/l, respectively. These levels of tolerance to kanamycin are consistent with those reported for selection purposes with agronomic species transformed using these vectors. Adventitious shoots of putative larch transformants were isolated after modified leaf disc incubation with *A. tumefaciens* or after hypocotyl inoculation with *A.*

M. R. Ahuja (ed.), Somatic Cell Genetics of Woody Plants, 55–63.

rhizogenes. These shoots are being grown for verification of transformation via Southern blotting. We believe our larch transformation system will serve as a valuable model system for gene transfer in other coniferous species.

2. INTRODUCTION

Larch species (*Larix* spp.) exhibit several characteristics which show them to be appropriate and excellent candidates for improvement via newly developing forest biotechnologies. First, unimproved and unselected European larch (*Larix decidua* Mill.) and Japanese larch (*L. leptolepis* [Sieb. and Zucc.] Gord. and their hybrids (*L. eurolepis* Henry) outgrow other conifers in height and volume in side-by-side tests in the northern United States and Canada (5,13,16,26,28,35,37,38). In addition to their superior growth characteristics, larches maintain a good wood density and have pulping yields and pulp strength characteristics similar to conifers twice their age (12). Second, larches are highly variable in form, growth rate, pest resistance, site requirements and cold hardiness, thus offering substantial opportunity for genetic improvement (6,15,24,32,36,39,43). Third, since larches generally show sporadic seedset, low seed yields and poor seed quality with 20-40% germination being typical, (4,17,18), propagation via asexual means such as cell and tissue culture would be beneficial. Fourth, larches are amenable to tissue culture propagation from juvenile (11,22) and mature tissues (3) and they are amenable to genetic manipulation using *Agrobacterium rhizogenes* and *A. tumefaciens* (9,10). Fifth, single gene traits for herbicide resistance (e.g. the glyphosate resistance genes) and insect resistance (e.g. the *Bacillus thuringiensis* toxin genes) incorporated in *Agrobacterium* plasmids could have a very significant effect in improving larches which are extremely sensitive to herbicides (33) and whose primary insect problems are those due to defoliating insects (39,40).

Thus, we believe that larches serve as an excellent model system for forest biotechnology research on conifers. In this paper, we briefly describe the methods for both larch micropropagation from juvenile tissue and for transformation of larches. Following, we discuss potential uses of these technologies for larch improvement.

3. MICROPROPAGATION

Adventitious bud production on juvenile tissues of European larch, Japanese larch, and tamarack (*Larix laricina* [Du Roi] K. Koch) is initiated by a one to two week treatment with 5 to 10 mg/l benzylamino purine (BAP). Buds are induced on cut cotyledon, hypocotyl or young needle segments (11,22) or on intact young seedlings inverted into initiation media (30)

The inverted seedling system, modified from that described

by Aitken-Christie, Singh, and Davies (2) for radiata pine, produces an average of about 70 adventitious buds per larch genotype. Clones of 250 or more have been produced using this system. In this process, seedlings inverted for 6 weeks (2 weeks initiation followed by 4 weeks post-initiation development on medium devoid of growth regulators) are returned to the upright position for bud elongation. The vitreous bud phenotype, a serious problem in our earlier work, was reduced in frequency by 80% through halving the nitrogen level in our elongation media (27).

Shoot elongation is effected on any of a number of media including those of Litvay (25), Gresshoff and Doy (29) and Le Poivre (34). We generally supplement our elongation media with 10.0 g/l sucrose and 10.0 g/l activated charcoal. Stimulation of shoot elongation, especially on shoots which have shown reduced elongation rates, can often be achieved by 2-week pulse treatments on media supplemented with 0.5 mg/l BAP. Both initiation and subsequent elongation of buds are supported by incubation in sealed petri dishes incubated at 20°C under constant 6500 lux cold fluorescent (70%) and incandecent (30%) illumination.

Rooting of larch adventitious shoots can either be done *in vitro* or in the greenhouse. The addition of 0.5 mg/l napthalene acetic acid (NAA) to the elongation medium will induce *in vitro* root formation within about 4 weeks. While root initiation is readily effected *in vitro*, the quality of the root system is generally less than that produced in soil mixes. Thus, we more commonly dip bases of our 2-3 cm long shoots in Rootone F and place them in pro-mix under intermittent mist in the greenhouse. Rooted plantlets are further grown in the greenhouse to a height of about 15 cm whereupon they are ready for field planting. We have established field trials comparing tissue culture plants and seedlings at several locations in the Upper Great Lakes region.

Remultiplication of tissue culture plants was carried out both *in vitro* and in the greenhouse. First, needles removed from adventitious shoots were subjected to initiation under previously described conditions (11). This may be done indefinitely through successive generations. Second, rooted tissue culture plants bent from their base were affixed horizontally after removal of the leader. This induced multiple axillary and adventitious buds to elongate which were then harvested and rooted essentially as cuttings (23). These two techniques offer good potential for scaling up the numbers of desirable genotypes.

4. GENE TRANSFER

Genetic engineering progress with conifers has been hampered by problems relating to the lack of infection of most conifers by Agrobacterium (8) and the fact that we cannot yet routinely regenerate whole plants from conifer cells or protoplasts that might be used for genetic manipulation (20,21). We have, however, been able to successfully infect young larch seedlings with both *Agrobacterium rhizogenes* and *A. tumefaciens*, and we are working on developing techniques for obtaining transformed plants via adventitious bud production.

We infected European larch seedlings with both *A. rhizogenes* strains and three of the six *A. tumefaciens* strains used as inocula. Site of inoculation was important for subsequent infection. Hypocotyl wounds were best for all strains tested as compared to cotyledon or root inoculations.

Our regeneration studies have focused on three factors: obtaining suitable adventitious organ formation following inoculation, using kanamycin as a screening agent for selecting transformed cells and organs, and using a modified "leaf disc" procedure for direct selection of transformed regenerants. Our most promising results have taken place following inoculation with *A. rhizogenes* ATCC11325 and pARC-8 (9). Inoculation with ATCC11325 has consistently resulted in multiple adventitious shoots being produced while pARC-8 inoculations have resulted in multiple root formation, typical of the hairy root disease. Both of these types of organs have been used for successfully regenerating whole transformed plants of agricultural crops (7,41,42). We have obtained typical *A. tumefaciens* galls with A208, pWB101/WAG11, and FACH. However, we were not successful in attempts to initiate adventitious buds from these gall tissues.

Our kanamycin screening studies have concentrated on larch inoculated with *A. tumefaciens* with pWB101/WAG11 and *A. rhizogenes* with pARC-8. These two plasmids carry the neomycin phosphotransferase (NPT) gene conveying resistance to kanamycin. Calli from seedlings inoculated 30-days earlier were exposed to kanamycin-supplemented media. Four callus lines of 26 tested after inoculation with pWB101/WAG11 were resistant to 100 mg/l kanamycin (See Figure 4 in Appendix). Control callus was killed by levels as low as 20 mg/l kanamycin. Several callus lines transformed by pARC-8 were resistant to 30 mg/l kanamycin.

We have tried two modifications of the "leaf disc" method for regenerating transformed plants as described by Horsch *et al.* (19). Young seedlings, as well as vitreous and normal larch tissue culture propagules were cut to segments and incubated for two days with pWB101/WAG11. Segments were placed on 50 mg/l kanamycin-supplemented medium for two weeks, following which time surviving tissues were challenged with 100 mg/l. None of 51 genotypes so treated survived.

In a further modification of the "leaf-disc" procedure, segments of young seedlings were incubated in suspension with pWB101/WAG11, then placed on 50 mg/l kanamycin-supplemented, bud-initiation medium (22). We have obtained clusters of buds apparently resistant to kanamycin using this technique.

The fact that our putative transformed larch callus is resistant to kanamycin is circumstantial evidence for transformation. We have isolated the pWB101/WAG11 and larch DNA in preparation for Southern blotting to verify the incorporation of the plasmid DNA into larch.

5. POTENTIAL FOR GENETIC IMPROVEMENT

We have successfully and repeatedly infected *Larix decidua* seedlings with several strains of *Agrobacterium rhizogenes* and *A. tumefaciens*. Symptom expression has been rapid ($\pm$ 1 month) and has occurred with high frequency. We have also successfully obtained expression of the bacterial NPT gene conveying resistance to kanamycin in our larch callus. We are attempting to develop a method for obtaining transformed plants using adventitious bud production following inoculation with *A. rhizogenes*.

The first practical application of this research could take two forms. First, larches are notoriously sensitive to herbicides. Therefore a suitable objective for a commercially important trait in larch would be to attempt to transfer in resistance to a herbicide such as glyphosate. Currently, there is no herbicide available for release of larches from competition as larches are very susceptible to herbicides (33). Successful transfer and expression of the ARO-A gene for glyphosate resistance has been obtained in *Populus* by Fillatti *et al*. (14) although the level of gene expression is still below that practical for commercial use. However, this pioneering research shows that trees will likely be amenable to the same gene transfer manipulations that are being used on agricultural crops.

A second potential short-term commercial application of genetic engineering technology would be to transfer a gene controlling production of the *Bacillus thuriengensis* (BT) toxin. This gene has been isolated and spliced into plasmids by co-author Barnes. The two most common and damaging insect problems of exotic larches planted in North American are the larch sawfly and the larch casebearer (39). It is likely that larvae of both these species would be less damaging to larch trees expressing even a low level of the BT toxin gene.

Once a commercially important gene has been introduced into larch, then methods for large scale propagation of these transformed genotypes must be developed. Our current micropropagation system offers sufficient multiplication potential for research-scale testing, but it is not yet cost-efficient for a commercial-scale venture. Thus, additional effort will be focused in our laboratory towards automating our micropropagation system. Hedging shoots *in vitro* as described by Aitken-Christie and Jones (1) and continous production of shoots by a modification of the inverted embryo system (2) both offer expanded possibilities for clonal propagation via adventitious bud production. In addition, Nagmani and Bonga (31) have demonstrated embryogenesis in subcultured callus of European larch indicating that this is another potential propagation method to examine.

6. ACKNOWLEDGEMENT

Research described in this paper was partially supported by grants from the USDA Competitive Grants Program and the Michigan Biotechnology Institute.

REFERENCES

1. Aitken-Christie, J. and C. Jones. 1987. Towards automation: Radiata pine shoot hedges *in vitro*. Plant Cell, Tissue and Organ Culture 8: 185-196.
2. Aitken-Christie, J., A.P. Singh, and H. Davies. 1987. Multiplication of meristematic tissue - A new tissue culture system for radiata pine. Proc. Inter. Symp. Genetic Manipulation of Woody Plants. Michigan State Univ. (In Press).
3. Bonga, J.M. 1984. Adventitious shoot and root formation in tissue cultures of mature *Larix decidua*. Proc. Int. Symp. of Recent Advances In Forest Biotech., Michigan Biotechnology Institute. East Lansing, MI. pp. 66-68.
4. Campbell, A.J. 1983. Nursery practice and early establishment of larch. In: Proc. Larch Symposium: Potential for the Future. University of Toronto, Toronto, Ontario. pp. 11-26.
5. Carter, K.K. and L.O. Selin. 1987. Larch plantation management in the Northeast. Northern J. Applied For. 4: 18-20.
6. Cook, D. 1969. Planted larch in New York. 116 pp.
7. David, C., M.-D. Chilton, and J. Tempe. 1984. Conservation of T-DNA in plants regenerated from hairy root cultures. Bio/Technology 2: 73-76.
8. De Cleene, M. and J. De Ley. 1979. The host range of crown hall. Bot. Rev. 42: 389-411.
9. Diner, A.M. and D.F. Karnosky. 1987a. Differential responses of two conifers to *in vitro* inoculation with *Agrobacterium rhizogenes*. Eur. J. For. Path. (In Press).
10. Diner, A.M. and D.F. Karnosky. 1987b. *Agrobacterium*-mediated gene transfer in European larch. Proc. Inter. Symp. Genetic Manipulation of Woody Plants. Michigan State University. (In Press).
11. Diner, A.M., A. Strickler, and D.F. Karnosky. 1987.

Initiation, elongation, and remultiplication of _Larix decidua_ micropropagules. New Zealand J. For. Res. (In Press).

12. Einspahr, D.W., T.D. McDonough, and T. Joachimedes. 1983. Kraft pulping characteristics of European, Japanese, and European X Japanese larch hybrids. TAPPI J. 66(8): 72-76.
13. Einspahr, D.W., G.W. Wychoff, and M. (Harder) Fiscus. 1984. Larch -- A fast-growing fiber source for the Lake States and Northeast. J. Forestry 83: 104-106.
14. Fillatti, J.J., B.H. McCown, J. Sellmer, and B. Haissig. 1986. The introduction and expression of a gene conferring tolerance to the herbicide glyphosate in _Populus_ NC5339. Proc. TAPPI Research and Development Conf. pp. 83-85.
15. Giertych, M. 1979. Summary of results on European larch (_Larix decidua_ Mill.) height growth in the IUFRO 1944 provenance experiment. Silvae Gen. 28: 244-256.
16. Hall, J.P. 1983. Comparison of the growth of larch and other conifers on reforested and afforested sites in Newfoundland. For. Chron. 59: 14-16.
17. Hall, J.P. and I.R. Brown. 1976. Microsporogenesis, pollination and potential yield of seed of _Larix_ in NE Scotland. Silvae. Gen. 25: 132-137.
18. Hall, J.P. and I.R. Brown. 1977. Embryo development and yield of seed in _Larix_. Silv. Gen. 26: 77-84.
19. Horsch, R.B., J.E. Fry, N.L. Hoffman, S.G. Rogers and R.T. Fraley. 1985. A simple and general method for transferring genes into plants. Science 227: 1229-1231.
20. Karnosky, D.F. 1981. Potential for forest tree improvement via tissue culture. BioScience 31: 114-120.
21. Karnosky, D.F. 1987. Regeneration of forest trees from tissue culture. Invited Paper. Proc. 1984 USFS Workshop of Genetic Engineering in Forestry. Placerville, CA (In Press).
22. Karnosky, D.F. and A.M. Diner. 1984. A cotyledon culture system for cloning _Larix decidua_ and _Pinus banksiana_. In: TAPPI Research and Development Conference. Atlanta, Georgia. pp. 13-15.
23. Karnosky, D.F. and A.M. Diner. 1986. Techniques for increasing clone sizes in _Larix_. Proc. Sixth Inter. Congress of Plant Tissue and Cell Culture. Univ. Minn. p. 10 (Abstract).
24. Keiding, H. and H.C. Olsen. 1965. Assessment of stem form in clones and progenies of larch. Silvae Gen. 14: 115-122.
25. Litvay, J.D., M.A. Johnson, D. Verma, D. Einspahr, and K. Weyrauch. 1981. Conifer suspension culture medium development using analytical data from developing seeds. Instit. Paper Chem. Tech. Pap. 115. 17 pp.
26. MacGillivay, H.G. 1969. Larches for reforestation and tree improvement in Eastern Canada. For. Chron. 45: 440-444.
27. McLaughlin, J. and D.F. Karnosky. 1987. Overcoming _in vitro_ vitrification through media manipulationf or larch.

Proc. Inter. Symp., Genetic Manipulation of Woody Plants. Michigan State Univ. (In Press).

28. Mead, D.A. 1978. Comparative height growth of eastern larch and black spruce in northwestern Ontario. For. Chron. 54: 296-297.
29. Mehra-Palta, A., R.H. Smeltzer, and R.L. Mott. 1978. Hormonal control of induced organogenesis: experiments with excised parts of loblolly pine. TAPPI 61: 37-40.
30. Mulcahey, A.T. and D.F. Karnosky. 1986. High frequency adventitious bud production in Larix decidua. Proc. Sixth Inter. Congress of Plant Tissue and Gell Culture. Univ. Minn. p. 245 (Abstract).
31. Nagmani, R. and J.M. Bonga. 1985. Embryogenesis in subcultured callus of Larix decidua. Can. J. For. Res. 15: 1088-1091.
32. Park, Y.S. and D.P. Fowler. 1983. A provenance test of Japanese larch in eastern Canada, including comparative data on European larch and tamarack. Silv. Genet. 32: 96-101.
33. Perala, D.A. 1982. Early release -- Current technology and conifer release. In: Proc. Artificial Regeneration of Conifers in the Upper Great Lakes Region. Michigan Technological University. Houghton, MI pp. 396-410.
34. Quoirin, M. and P. Le Poivre. 1977. Etude de milieux adaptes aux cultures 'in vitro' de Prunus. Acta Hort. 78: 437-442.
35. Rauter, R.M. and B.J. Graham. 1983. The genetics and tree improvement potential of larches in Ontario. Proc. Larch Symp.: Potential for the Future. Ontario Ministry of Natural Resources. pp. 11-26.
36. Reck, S. 1977. Ergebnisse einer Versuchsanlage mit europaischen Larchen (Larix decidua Mill.) und Hybridlarchen (Larix eurolepis Henry). Silvae Gen. 26: 95-101.
37. Reed, D.D., R.L. Sajdak, and J. Kotar. 1983. Growth comparisons of tamarack and European larch in Upper Michigan. Proc. 3rd. North Central Tree Improv. Conf. pp. 71-77.
38. Reimenschneider, D.E. and H. Nienstaedt. 1983. Height growth to age 8 of larch species and hybrids in Wisconsin. USDA Forest Service Res. Pap. NC-239. 6 pp.
39. Robbins, K. 1985. Risks associated with growing non-native larches in eastern North America. North J. Appl. For. 2: 101-104.
40. Ryan, R.B., S. Tunnock, and F.W. Ebel. 1987. The larch casebearer in North America. J. Forestry 85(7): 33-39.
41. Shah, D.M., R.B. Horsch, H.J. Klee, G.M. Klee, G.M. Kishore, J.A. Winter, N.E. Tumer, C.M. Hironaka, P.S. Saunders, C.S. Gasser, S. Aykent, N.R. Siegel, S.G. Rogers and R.T. Fraley. 1986. Engineering herbicide tolerance in transgenic plants. Science 233: 478-481.
42. Shahin, E.A., K. Sukhapinda, R.B. Simpson and R. Spivey.

1986. Transformation of cultivated tomato by a binary vector in *Agrobacterium rhizogenes*: transgenic plants with normal phenotypes harbor binary vector T-DNA, but no Ri-plasmid T-DNA. Theor. Appl. Genet. 72: 770-777.

43. Toda, R. and S. Mikami. 1976. The provenance trials of Japanese larch established in Japan and the tentative achievements. Silv. Genet. 25: 209-216.

REGENERATION AND TRANSFORMATION OF APPLE PLANTS USING WILD-TYPE AND ENGINEERED PLASMIDS IN AGROBACTERIUM SPP.

D.J. JAMES, A.J. PASSEY, Institute of Horticultural Research, East Malling, Maidstone, Kent, ME19 6BJ, UK.
S. PREDIERI, Consiglio Nazionale delle Richerche, Via Filipo Re. 6, 40126 Bologna, Italy.
E. RUGINI, Consiglio Nazionale delle Richerche, Centro di Studio per la Olivicoltura, Borgo XX Giugno, 06100 Perugia, Italy.

ABSTRACT

We have examined a number of variables that control the frequency of shoot regeneration from apple leaf discs and pieces. These have included chemical factors such as auxin: cytokinin concentrations, the presence of putrescine and physiological factors such as orientation of leaf piece and stage of leaf expansion before excision.

Under optimal conditions shoot regeneration from apple leaf tissues approaches 100% with several shoots per disc. Using these conditions transformed apple callus and shoots have been produced using the leaf disc method after infection with disarmed strains of *Agrobacterium tumefaciens* that carry the engineered binary vector, Bin 6. Transformation has been shown by the presence of nopaline in leaf tissues and ability of the plants to root in the presence of kanamycin. Further molecular evidence is being sought.

Wild-type and engineered plasmids of *A. rhizogenes* have also been used in attempts to regenerate both genetic chimaeras and wholly transformed plants of several fruit tree subjects.

1. INTRODUCTION

Temperate fruit trees have yielded to improvement by conventional breeding methods over many decades and both rootstock and scion varieties have been bred for a great variety of horticultural characteristics. At the same time a vast amount of information has accrued relating to performance and orchard performance and appropriate cultivars can be selected for different climatic and soil types (e.g. 1). Few would argue however that there is not room for improvement in many commercial cultivars in terms of rootiness, dwarfing capacity, disease resistance, cold tolerance, drought resistance etc. It is in adding these 'fine' adjustments to existing commercially important cultivars that the newly emerging science of somatic cell genetics, coupled with powerful tissue culture methodology will make a great impact in fruit tree improvement in the future (2).

As with all plant subjects which one is aiming to improve by novel somatic methods it is necessary to develop reproducible *in vitro* regeneration procedures for cell and tissue cultures. The main aim of our work is to develop efficient regeneration procedures for fruit tree scions and rootstocks so that the twin methodologies of somaclonal variation (3) and transformation via *Agrobacterium* species can be exploited. These two methodologies are inextricably linked since it is necessary to regenerate from cells and tissues in both cases.

M. R. Ahuja (ed.), Somatic Cell Genetics of Woody Plants, 65–71.

Regenerating whole plants from complex explants such as leaf or root pieces has led to somaclonal variation i.e. variations in phenotype and genotype of regenerated plants. In many crop species improved variants have already been found but not surprisingly there are as yet no definitive reports of this phenomenon in trees although in cherry plants regenerated from root callus (4) there were no obvious signs of somaclonal variation in young potted plants.

An understanding of the factors controlling the regeneration process is vital in developing genetic manipulation techniques for tree species. The following paper describes our work on the factors controlling regeneration from apple leaf tissues and how this has been used in transformation studies using both wild-type and engineered plasmids in *Agrobacterium tumefaciens* and *A. rhizogenes*.

2. PROCEDURE

2.1. Materials and methods

2.1.1. *Micropropagation and regeneration*. Shoot cultures of the apple cultivars Greensleeves, Tuscan, M.9, and M.25 (*Malus pumila* Mill.) were propagated *in vitro* as previously described (5) except that the basal medium Murashige and Skoog (1962) (6) replaced the Linsmaier and Skoog (1965) medium (7) used in earlier work. Putrescine and 5-azacytidine were added to the medium by filter-sterilisation (Millipore 0.22 um). The growth regulators 6-benzylaminopurine (BAP) and 1-naphthalene acetic acid (NAA) were added before autoclaving at concentrations of 1 mg/l and 0.1 mg/l respectively. In other experiments these concentrations were varied from 0.1-0.5 mg/l NAA and 1.0-5.0 mg/l BAP.

Leaf discs were excised aseptically using a 7 mm cork borer and cultured in multiwell dishes (Sterilin) as previously described (8). All excised leaf material was grown in the dark at 25°C and regenerated shoots recorded aseptically at intervals up to 60 days after excision using a low power binocular microscope.

The variables examined for effects on regeneration from leaf explants using the cultivar Greensleeves were: 1) Effects of putrescine (1.6, 16.0, 161.0 mg/l) and 5-azacytidine (0.7, 7.0 mg/l) and their interaction with leaf disc origin and orientation of disc (abaxial or adaxial surface down); 2) Repeated regeneration of discs; 3) Auxin (NAA, 0.1, 0.5 mg/l)and cytokinin (BA, 1.0, 2.0, 5.0 mg/l) concentration; 4) Mode of excision; 5) X-irradiation.

Regenerated shoots still attached to leaf pieces were transferred to the light for shoot proliferation and subsequently for root induction using standard procedures (5).

2.1.2. *Transformation procedures*. Transformations were carried out using either the leaf disc method of Horsch *et al*. (9) by culturing leaf pieces in bacterial suspension of *Agrobacterium tumefaciens* or *A. rhizogenes* or by the shoot inoculation method of Ooms *et al*. (10). The bacterial strains used were (a) *Agrobacterium tumefaciens* carrying the disarmed Binary Vector Bin 6 (11) and (b) *A.rhizogenes*, wild-type agropine-producing strain LBA 9402, the kanamycin resistant derivative obtained by a triparental mating and named by us as

LBA 9402/19, a wild-type mannopine-producing strain carrying the plasmid pRi1896 and strain A4 Rs carrying the plasmid pRiB1017 conferring multiple antibiotic resistance (kanamycin, rifampycin and spectinomycin).

2.2. Statistical treatments

Both the percentage of leaf pieces that regenerated adventitious shoots and the number of shoots per regenerating disc were recorded. Percentage differences were examined using the normal approximation to a binomial distribution whilst data for shoot number per regenerating disc were subjected to analyses of variance using a Genstat 5 programme. Differences in rooting ability induced by the different strains of *A. rhizogenes* were analysed by a single factor ANOVAR.

3. RESULTS AND DISCUSSION

3.1. Regeneration from complex explants - somaclonal variation

Shoots arise in one of two ways on leaf explants. Those differentiating within the first 10-14 days appear mostly in association with veins and without visible callus formation. These presumably have an epidermal or subepidermal origin. Other shoots arise much later in association with callus usually forming at the cut edges of the leaf. These two populations have been separated and micropropagated independently.

Several factors affected regeneration frequency from leaf explants (14). Main effects were due to the auxin:cytokinin concentrations used. A range of media containing 1-naphthalene NAA and BAP gave regeneration frequencies that varied from 20-90% depending on auxin:cytokinin ratios and a series of other variables. For example, neither the diamine putrescine known to promote apple cell division (12) or 5-azacytidine known to affect cell differentiation (13) significantly affected mean shoot number compared with regeneration media lacking these compounds. This was true regardless of whether the leaf disc was placed with its adaxial or abaxial surface adjacent to the medium. However, when each of the individual treatments was compared with respect to leaf orientation there were significant effects on shoot number but not on percentage regeneration.

Using excised leaves from rooted shoots as opposed to unrooted shoots caused a significant increase in the percentage of discs regenerating only when 0.1 or 1.0 mM putrescine was present in the regeneration medium. There was no significant effect on mean shoot number per explant in any of the other treatments.

The effect of recurrently regenerating shoots over two or more cycles was examined by comparing regeneration frequencies with those discs derived from the parent clone. This was done to examine the possibility that one could select for cells competent to undergo organogenesis and successively raise the level of regeneration. Using such methods, however, failed to raise regeneration frequencies above those observed in control leaves excised from cultures that had never undergone any regeneration.

The importance of the manner of excising individual leaves was assessed by comparing 7 mm discs and the leaf remainder with leaves cut into 3 or more strips. Data were expressed on per leaf basis rather than a 'per excised explant' basis. The main conclusion from this work showed that cutting the leaf into 3 or more pieces significantly enhanced shoot number per leaf regardless of whether the cuts were made parallel or transversely to the long axis when compared to cutting discs.

A range of scions and rootstocks have now been regenerated using this information including the commercially important rootstocks M.9, M.25, (14) M.26 and M.27 (8) and the compact columnar 'Wijcik' types (14, 15). The frequency of regeneration in other cultivars varied considerably but was generally in the range 30-60%. In the cultivar Greensleeves distinction has been made between shoots regenerated with and without callus formation amongst the regenerated plants transferred to orchard trials for assessment of somaclonal variation.

3.2.Transformation with Agrobacterium spp.

Wild-type strains of *Agrobacterium tumefaciens* and *Agrobacterium rhizogenes* are being used in attempts to transform different fruit tree scions and rootstocks for 'shootiness' and 'rootiness'.

3.2.1.Agrobacterium tumefaciens - Wild type plasmids.

Work with potato (10) has shown there were marked effects on the reproductive physiology of plants regenerated from galls originally induced by inoculation of micropropagated shoots. These Ti-transformed plants showed fundamental differences in patterns of fruit set and abscission. Since these processes are of great importance in fruit trees the existence of Ti-transformed fruit trees could provide valuable insights into their basic control mechanisms.

Preliminary experiments with a wild-type strain of *Agrobacterium tumefaciens* harbouring the plasmid pTiT37 has shown that nopaline-type galls regenerate shoots quite readily in cherry and less so in apple and pear (16). Although plants could be successfully transplanted to pots, when tested for nopaline they gave a negative result suggesting selection for untransformed cells and cross-feeding from transformed cells. The fact that they could be induced to root *in vitro* for soil transfer and had a normal phenotype further suggests that they were untransformed regenerates.

3.2.2. *Agrobacterium tumefaciens - Disarmed, engineered plasmids - Binary Vectors*. An additional approach is to use strains of *Agrobacterium tumefaciens* that contain engineered 'binary' vectors. These have had the genes coding for auxin and cytokinin deleted so that they do not interfere with the regeneration process. Such vectors are said to be 'disarmed' and transformed regenerated plants should be morphologically noraml in appearance. We have used the binary vector Bin 6 (11) for transforming apple leaf discs of the cultivars Greensleeves and Tuscan (17). This 'disarmed' vector is a chimaeric gene construct composed of the T-DNA border sequences, the regulating sequences of an opine (nopaline) synthase gene and the coding sequences of a bacterial neomycin phosphotransferase gene conferring resistance to the antibiotic kanamycin (kan+).

Transformed plants thus show growth on kanamycin-containing media whilst non-transformed plants 'bleach' and die. The ability to root on kanamycin is usually taken as a strong indication that transformation has occurred. Usually transformants also contain detectable levels of nopaline. Both genes can be used as dominant selectable markers. Initial experiments showed that rooting percentages of untransformed apple shoots was zero in the presence of more than 50 mg/l kanamycin. We have noted over several years that non-transformed apple shoots regenerated from leaf pieces can sometimes show a permanent loss or large reduction in rooting ability *in vitro* compared to the parent lines.

Transformed apple callus has been obtained in co-culturing experiments (17) where selection for kan+ cells is made immediately after transformation. In twelve experiments about 35% of leaf discs formed callus in the continued presence of 100 mg/l kanamycin. These calli were positive for nopaline synthase when tested using paper electrophoresis (18) whereas control disarmed plasmids (LBA 4404) lacking the kanamycin gene gave no callus. The kanamycin resistant callus has now regenerated shoots albeit at a very low frequency in a cultivar in which up to 90% regeneration can be obtained from untransformed tissues. The transformed shoots took almost 5 months to regenerate compared to the 10-20 days normally taken when using untransformed discs. If selection is made for only a limited period after transformation (5-10 days) and then the discs transferred to an optimal regeneration medium without kanamycin many shoots regenerate but of those so far tested (ca. 30) none contained nopaline. If the interval between infection and challenge was reduced to three days regeneration fell to 8%. In other experiments apple leaf discs were infected with Bin 6 and allowed to regenerate for two months without kanamycin and then transferred to a shoot proliferation medium containing 100 mg/ml kanamycin. About 16% of these shoots remained green after 28 d exposure.

3.2.3. Agrobacterium rhizogenes. Co-culture. A range of apple cultivars could be induced to form roots directly on wounded leaf margins after co-culture with bacterial suspensions of the wild-type strain LBA 9402 and the kanamycin resistant derivative LBA 9402/19. In preliminary experiments the frequency of root formation on discs was 10-20% but more recent experiments have given 100% rooting. In addition roots have been produced on leaf discs after co-culturing experiments and then shoots regenerated from these roots. These shoots then produced roots on cytokinin-based media normally inducing shoot proliferation. Copious root production was accompanied by a total lack of axillary shoot proliferation, suggesting they have the Ri phenotype (19). These plants await agropine testing.

3.2.4. Shoot inoculation. In shoot injection experiments the percentage of M.9 apple roots that formed at the site of infection after inoculation with strain LBA 9402 was influenced by temperature during the culture period after infection. At 29°C the percentage of shoots forming roots after approximately one month was usually about 80 whilst at 25°C this was never more than 25, an effect observed in four separate experiments.

Plants produced in these experiments are viable genetic chimaeras possessing an untransformed shoot growing with transformed roots. The phenotype of the Ri transformed root system would be expected to have different physiological, biochemical and physical properties compared to untransformed adventitious roots (19). Several rootstocks have or are being produced with this character and will shortly enter trials.

Observations and measurements have been made on the differences in growth patterns and rates between transformed and non-transformed root systems of M.9 apple rootstocks. *In vitro* root systems initiated at the same time and grown for one month showed that increases in fresh weight of the untransformed root system were significantly greater than any of the three different root systems induced by strains LBA 9402, 9402/19 and A4 Rs. There were no significant differences between the fresh weights of any of the putatively transformed root systems.

Qualitative differences in the root systems were observed when the chimaeras were transferred to soil and allowed to grow under greenhouse conditions for three months. When plants were removed from the pots all putatively transformed root systems were much more fibrous than control non-transformed roots and showed much more branching. This was testified by the greater difficulty encountered in removing soil from the transformed root systems. Tests are being carried out to examine putatively transformed roots for the presence of opines.

Work has continued to regenerate shoots from transformed roots and several new media and growth conditions tried but so far without success. A different approach has recently been used that involves excision of the infected region at different times following infection and exposure to regeneration media before differentiation of transformed roots. This has produced shoots in some cases and these are being grown on for assessment.

4. ACKNOWLEDGEMENTS

We thank the following for kindly supplying us with the various strains of Agrobacteria and their plasmids. Mike Bevan (PBI, Cambridge, UK) Bin 6; Gert Ooms (Rothamsted, Harpenden, UK) T37, LBA 9402; John Hammill (FRI, Norwich, UK) LBA 9402/19; Jacques Tempe (Orsay, Paris, France) pRi1896; Francine Casse-Delbart (INRA, Paris, France) A4 Rs.

REFERENCES

1. Brown AG: Apple. In: Advances in Fruit Breeding (J Janick and JN Moore Eds), pp.3-37. Purdue University Press, West Lafayette, Indiana. 1975.
2. James DJ: Cell and tissue culture technology for the genetic manipulation of temperate fruit trees. In: Biotechnology and Genetic Engineering Reviews, vol. 5. Intercept, Newcastle-upon-Tyne, pp.33-79. 1987.
3. Larkin PJ: Scowcroft WR: Somaclonal variation: a novel source of variability from cell cultures for plant improvement. Theor. Appl. Genet. 60, pp.197-214. 1981.

4. James, DJ, MacKenzie KAD, Malhotra SB: The induction of hexaploidy in cherry rootstocks using in vitro regeneration techniques. Theor. Appl. Genet., 73, 589-594. 1987.
5. James DJ, Thurbon IJ: Rapid in vitro rooting of the apple rootstock M.9. J. Hort. Sci., 54, 309-311. 1979.
6. Murashige T, Skoog F: A revised medium for rapid growth and bioassays with tobacco tissue cultures. Physiologia Plantarum, 15, 473-497. 1962.
7. Linsmaier EM, Skoog F: Organic growth factor requirements of tobacco tissue culture. Physiol. Plant. 18, 100-127. 1965.
8. James DJ, Passey AJ, Malhotra SB: Organogenesis in callus derived from stem and leaf tissues of apple and cherry rootstocks. Plant Cell Tissue Organ Culture, 3, 333-341. 1984.
9. Horsch RB, Fry JE, Hoffman NL, Eicholtz DA, Rogers SG, Fraley RT: A simple and general method for transferring genes into plants. Science, 227, 1229-1231. 1985.
10. Ooms G, Karp A, Roberts J: From tumour to tuber; tumour cell characteristics and chromosome numbers of crown gall-derived tetraploid potato plants (Solanum tuberosum cv. Maris Bard). Theor. and Appl. Genet., 66, 169-172. 1983.
11. Bevan M: Binary Agrobacterium vectors for plant transformation. Nucl. Acids Res. 12, pp.8711-8721. 1984.
12. James DJ, Passey AJ, MacKenzie KAD, Menhinick EC: Protoplast culture and the use of apomixis in vitro for the regeneration of fruit plants. In: Plant tissue culture and its agricultural applications. (LA Withers and PG Alderson, Eds) pp.349-359. Butterworths, London. 1986.
13. Hepburn AG, Clarke LE, Pearson L, White J: The role of cytosine methylation in the control of nopaline synthase gene expression in a plant tumor. J. Mol. Applied Genet. 2, pp.315-329. 1983.
14. James DG, Passey AJ, Rugini E: Factors affecting high frequency plant regeneration from apple leaf tissues cultured in vitro. J. Pl. Physiol. (In Press). 1987.
15. Lapins KO: Segregation of compact growth types in certain apple seedling progenies. Can. J. Plant Sci. 49, 765-768. 1969.
16. James DJ, Murphy H. Passey AJ: Transformation of apple, cherry and plum rootstocks. In: Rep. East Malling Res. Stat for 1984, p.118. 1985.
17. James DJ, Passey AJ: Binary vectors. In: Rep. East Malling Res. Stat for 1985. p.95. 1986.
18. Otten LABM, Schilperoort RA: A rapid microscale method for the detection of lysopine and nopaline dehydrogenase activities. Biochem. Biophys. Acta, 527, pp.497-500. 1978.
19. Tepfer D: The biology of genetic transformation of higher plants by Agrobacterium rhizogenes. In: Molecular Genetics of the bacteria-plant interaction. (A Puhler, Ed). pp.248-258. Springer-Verlag, Berlin, Heidelberg. 1983.

EXPRESSION OF AN HERBICIDE TOLERANCE GENE IN YOUNG PLANTS OF A TRANSGENIC HYBRID POPLAR CLONE

D.E. RIEMENSCHNEIDER[1], B.E. HAISSIG[1], J. SELLMER[2], AND J.J. FILLATTI[3]

ABSTRACT

The herbicide glyphosate (N-phosphono methylglycine) kills plants primarily by inhibiting the enzyme EPSP synthase (5-enolpyruvylshikimate-3-phosphate synthase). The *aroA* gene codes for EPSP synthase. An altered *Salmonella typhimurium aroA* gene, which codes for a less susceptible EPSP synthase, was previously engineered for expression in higher plants, and inserted into hybrid poplar NC-5339 cv. 'Crandon' (*Populus alba* x *P. grandidentata*) by using an *Agrobacterium tumefaciens* binary vector.

In the present tests, plants of such genetically transformed (subclones) and untransformed (control, parent) clones were propagated *in vitro* by continuous shoot culture and transplanted to the greenhouse. After 4 weeks in the greenhouse, plants were sprayed with 0.07 and 0.28 kg/ha (active ingredient) of a commercial formulation of glyphosate (Roundup, Monsanto Chemical Co.). Height growth was measured at 3-d intervals for 15 d.

Results indicated that the growth rate of five of the six subclones tested was significantly greater than plants of the untransformed parent clone after treatment with 0.07 kg/ha Roundup. Plants of one sublcone grew an average of 10.3 cm during 15 d after treatment with 0.07 kg/ha Roundup. In comparison, identically treated plants of the untransformed parent clone grew only 4.3 cm during the same period. When plants were sprayed with 0.28 kg/ha Roundup, plants of subclones grew significantly taller than those of the untransformed parent clone, but the height differences were less than at the lower dosage.

1. INTRODUCTION

Genetic transformation of trees has been accomplished by using the natural infection mechanism of *Agrobacterium tumefaciens*, the bacterium causing crown gall disease (3, 5, 6, 7, 12, 13). In this method, desired genes are incorporated into the bacterial T-DNA, then transferred to the higher plant during the bacterial infection (8).

---

[1] USDA Forest Service, North Central Forest Experiment Station, Forestry Sciences Laboratory, Rhinelander, WI 54501, USA.

[2] University of Wisconsin, Department of Horticulture, Madison, WI, 53706, USA.

[3] Calgene, Inc., 1920 Fifth Street, Davis, CA, 95616, USA.

M. R. Ahuja (ed.), Somatic Cell Genetics of Woody Plants, 73–80.

Our long-term objective is to genetically engineer hybrid poplars that resist the herbicide glyphosate (N-phosphono methylglycine). Glyphosate kills plants primarily by inhibiting the enzyme EPSP synthase (5-enolpyruvylshikimate-3-phosphate synthase, EC 2.5.1.19) (16). An altered *aroA* gene that codes for a glyphosate tolerant EPSP synthase (2, 15) was engineered for expression in plants. We introduced that gene into hybrid poplar NC-5339 (*Populus alba* x *P. grandidentata* cv. 'Crandon'), using a binary oncogenic strain of *A. tumefaciens* as a vector (5, 6), to produce genetically transformed "subclones" of NC-5339. Laboratory tests have verified the incorporation and expression of the altered *aroA* gene in some subclones of this hybrid poplar (6 and Table 1).

TABLE 1. Identities of six subclones of the parent clone (NC-5339). The subclones contained the altered *aroA* gene and were tested for tolerance to Roundup in a greenhouse (see text for detail, see (6) for transformation procedures).

Transformed Subclone[1]	Alternate Identification	By Source	Method of Verifying Transformation
NC-12771	75-12	UW[2]	Western blot
NC-12778	LAE-2	"	NPT II' assay
NC-12797	39-122	Calgene[3]	Southern blot
NC-12804	39-115-H	"	NPT II' assay
NC-12809	75-12	UW	Southern and Western blot
NC-12812	39-112-A	Calgene	NPT II' assay

[1]North Central Forest Experiment Station accession number. The untransformed parent clone was included in all tests.

[2]University of Wisconsin-Madison, Department of Horticulture.

[3]Calgene, Inc., Davis, CA.

The present paper reports results of preliminary greenhouse testing of these transgenic NC-5339 subclones for tolerance to the herbicide Roundup[1] (Monsanto Chemical Co.,), a proprietary formulation containing glyphosate. It also reports results of testing these subclones for residual *Agrobacterium* contamination.

2. MATERIALS AND METHODS

2.1 Genetic Transformation

Genetic transformations were conducted at Calgene Inc., Davis, CA (6). The nopaline-producing strain of *Agrobacterium tumefaciens*, C58/587/85, was used as the transformation vector. This strain contains a wild-type, oncogenic Ti plasmid and a small plasmid with three chimeric genes: two of

---

[1] Mention of tradenames does not constitute endorsement by the United States Department of Agriculture or the Forest Service.

these genes coded for neomycin phospho-transferase (NPT II'), a selectable marker; the third was the altered *aroA* gene, fused to the mannopine synthase promoter (2). Leaf pieces taken from NC-5339 shoot cultures were cocultivated with *Agrobacterium* strain C58/587/85, and shoots were regenerated on selective media containing 60 mg/l of kanamycin (6).

Each shoot that regenerated after cocultivation with *Agrobacterium tumefaciens* strain C58/587/85 was numerically identified and maintained separately as an individual "subclone" during subsequent propagation and testing. Regenerated shoots (subclones) of normal morphology that exhibited NPT II' enzyme activity were transferred as *in vitro* cultures to the Forestry Sciences Laboratory, Rhinelander, WI, in March, 1986, for propagation of plants and greenhouse herbicide spray testing (Table 1).

2.2 *In Vitro Propagation*

Continuous shoot cultures (1C) of the six subclones (Table 1) and the parent clone, NC-5339, were propagated *in vitro* at 25 °C under Sylvania F72T12/CW fluorescent lamps (35 umol $m^{-2}s^{-1}$) on Murashige and Skoog medium containing sucrose (30 g/l), benzyladenine (0.1 mg/l), and Difco Bacto Agar (0.6% w/v): pH adjusted to 5.7 with 1.0 mol/l NaOH before autoclaving (15 m, 121 °C). Cultures were maintained in Magenta GA 7 (Magenta Corp., Chicago, IL) culture vessels containing 50 ml of medium, and subcultured at 30-d intervals. To propagate greenhouse plants, shoots were cultured for 30 d, shortened to 3 to 4 cm long, and rooted without chemical treatment in covered aluminum trays containing sterile potting medium (Redi-earth Peat-Lite Mix, W.R. Grace and Co., Cambridge, MA). The potting medium had previously been watered, then drained to field capacity. Rooting trays were maintained in the culture room for 14 d, after which the rooted plantlets were removed from the trays, repotted in 10.2 cm diameter plastic pots using the same medium, and placed on the greenhouse bench. Plants were grown in the greenhouse for an additional 28 d (24 °C days, 20 °C nights, 18 h photoperiod), at which time the plants were about 25 cm tall and had five to six well-developed leaves. These plants were used for Roundup spray tests.

2.3 *Roundup Spray Tests*

Plants of subclones and the parent clone were treated with Roundup by spraying them with a calibrated, overhead, moving boom-sprayer that we constructed. Dosages were none (control), and 0.07 and 0.28 kg/ha of active ingredient (isopropylamine salt of glyphosate). Ten to eighteen plants per clone per treatment were used in three replications in time. After leaf surfaces had air dried, plants were returned to the greenhouse bench for the remainder of the test.

Plant stem height was measured to the nearest 0.1 cm on the day of spraying (day 0) and every 3 d thereafter for 15 d. Net height growth was used as the basis for analysis and was calculated by subtracting height at day 0 from subsequent height measurements.

Data were analyzed by analysis of variance (ANOVA). First, the effect of herbicide treatment on height growth of all plants was tested according to the model:

$$H = u + T_i + R_k + TR_{ik} + e$$

where H is plant height, u is the population mean, T is the effect of spray treatment, R is the replication effect, TR is the treatment x replication interaction, and e is experimental error. Differences between treatment means at each measurement date were tested using single degree of freedom ANOVA comparisons.

Second, height growth data from each herbicide treatment were analyzed separately to test for differences between subclone and parent clone, according to the model:

$$H = u + C_j + R_k + CR_{jk} + e$$

where H, u, and e are as before; C is the subclone effect, R is the replication effect, and CR is the subclone x replication interaction.

Subclone means at the same or different treatment levels were compared using the least significant difference test.

2.4 Agrobacterium Assay

One shoot was randomly selected from each of 26 transformed subclones (including those in Table 1). Each shoot was about 5 cm long with three well-developed leaves. Shoots were divided into basal, second, third, and apical stem sections. Each section contained a leaf except for the apical section that contained the developing terminal leaf cluster. Sections were plated on media containing yeast extract (Difco, 1% w/v), glucose (2% w/v), calcium carbonate (2% w/v), and agar (Difco, 2% w/v). These plates were incubated at 25 °C for 48 h, then flooded with a shallow layer of Benedict reagent (1). Plates were then examined for yellow color. The test was replicated three times for each bacterial strain; *Agrobacterium* strain C58, which gives a positive reaction in the test, was included for comparison with unknown strains.

3. RESULTS

3.1 Roundup Spray Tests

Mean height growth of plants of untreated subclones and the untreated parent clone were equal (Table 2); plants of the subclones were also morphologically identical to the parent clone.

Overall, when plants were treated with 0.07 kg/ha Roundup, there was significantly less mean height growth within 9 d after spraying as compared to untreated plants. Treatment with 0.07 kg/ha Roundup resulted in significantly less mean height growth of plants of two of the six subclones (NC-12089 and NC-12812) and the parent clone, compared to untreated plants. However, height growth of plants of the other four subclones treated with 0.07 kg/ha was statistically equal to untreated plants by the end of the test (day 15, Fig. 1). At the end of the test, mean height growth of plants of all subclones treated with 0.07 kg/ha, except NC-12812, was significantly greater than plants of the identically treated parent clone (Fig. 1). Transformed plants frequently survived treatment with 0.07 kg/ha Roundup, but prolific auxillary shooting and poor plant form commonly resulted.

Based on analysis of pooled means for plants of all subclones and the parent clone, treatment with 0.28 kg/ha Roundup resulted in significantly less mean height growth within 3 d after spraying as compared to untreated plants. Height growth of plants of the subclones also differed from the untransformed parent clone after treatment with 0.28 kg/ha Roundup, although the differences were less significant than with the lower dosage. For example, 15 d after treatment with 0.28 kg/ha Roundup, the parent clone had grown, on average, 2.1 cm, compared to a mean of 3.4 cm for the most tolerant subclone (Fig. 2). Comparable mean measurements at the lower dosage (0.07 kg/ha) were 4.7 and 10.3 cm, respectively (Table 2). Treatment with 0.28 kg/ha Roundup ultimately killed all plants.

TABLE 2. Mean net height (±SE) of plants of subclones and their parent clone (NE-5339) at 3 d intervals after being sprayed with different dosages of Roundup. Differences between subclones and the parent clone were significant after treatment with Roundup at 0.07 kg/ha (P <0.01) and 0.28 kg/ha (0.01 <p <0.05). Differences between untreated subclones and the parent clone were not significant.

		Height Growth (cm) By Day After Treatment				
Dosage[1]	Subclone	3	6	9	12	15
0	NC-12271	1.8 ± 0.1	5.1 ± 0.6	7.4 ± 0.5	10.6 ± 1.3	12.3 ± 1.1
0	NC-12778	2.2 ± 0.2	5.3 ± 0.6	7.3 ± 0.6	10.7 ± 1.3	12.5 ± 1.1
0	NC-12797	1.7 ± 0.1	4.8 ± 0.4	6.8 ± 0.4	10.0 ± 0.8	11.8 ± 0.7
0	NC-12804	1.6 ± 0.2	4.9 ± 0.8	6.9 ± 0.7	10.3 ± 0.8	11.7 ± 0.5
0	NC-12809	1.9 ± 0.1	4.7 ± 0.6	6.8 ± 0.4	9.5 ± 1.4	11.5 ± 0.8
0	NC-12812	1.8 ± 0.1	5.0 ± 0.6	7.0 ± 0.5	10.6 ± 0.3	12.2 ± 1.0
0	NC-5339	1.9 ± 0.3	4.7 ± 0.8	6.9 ± 0.7	9.4 ± 1.1	11.3 ± 0.9
0	Mean	1.8 ± 0.1	4.9 ± 0.2	7.0 ± 0.2	10.2 ± 0.4	11.9 ± 0.4
0.07	NC-12771	2.0 ± 0.4	5.8 ± 1.4	6.9 ± 1.4	9.7 ± 2.4	10.3 ± 2.3
0.07	NC-12778	1.6 ± 0.4	4.1 ± 0.3	5.0 ± 0.2	6.8 ± 2.0	7.0 ± 1.9
0.07	NC-12797	1.7 ± 0.4	4.9 ± 1.3	5.9 ± 1.3	9.0 ± 2.3	9.5 ± 2.2
0.07	NC-12804	1.9 ± 0.6	5.8 ± 2.2	6.7 ± 2.2	9.4 ± 3.6	9.7 ± 3.5
0.07	NC-12809	1.3 ± 0.3	3.8 ± 1.2	4.8 ± 1.3	7.4 ± 2.7	7.8 ± 2.6
0.07	NC-12812	1.6 ± 0.4	3.9 ± 0.8	4.4 ± 0.7	5.6 ± 1.0	5.7 ± 1.0
0.07	NC-5339	1.0 ± 0.3	3.0 ± 0.7	3.5 ± 0.6	4.3 ± 0.8	4.7 ± 0.8
0.07	Mean	1.6 ± 0.2	4.5 ± 0.5	5.4 ± 0.5	7.6 ± 0.8	7.9 ± 0.8
0.28	NC-12771	0.8 ± 0.1	2.0 ± 0.3	2.4 ± 0.2	2.8 ± 0.3	2.9 ± 0.3
0.28	NC-12778	1.2 ± 0.3	2.4 ± 0.5	2.7 ± 0.5	3.2 ± 0.7	3.2 ± 0.7
0.28	NC-12797	0.7 ± 0.1	1.5 ± 0.1	1.9 ± 0.1	2.3 ± 0.2	2.4 ± 0.2
0.28	NC-12804	1.1 ± 0.3	2.4 ± 0.5	2.8 ± 0.5	3.5 ± 0.7	3.4 ± 0.7
0.28	NC-12809	0.5 ± 0.1	1.6 ± 0.3	1.8 ± 0.3	2.3 ± 0.5	2.4 ± 0.5
0.28	NC-12812	0.7 ± 0.1	1.5 ± 0.2	1.9 ± 0.2	2.0 ± 0.2	2.1 ± 0.2
0.28	NC-5339	0.6 ± 0.1	1.3 ± 0.2	1.7 ± 0.2	2.0 ± 0.3	2.1 ± 0.3
0.28	Mean	0.8 ± 0.1	1.8 ± 0.1	2.1 ± 0.1	2.6 ± 0.2	2.6 ± 0.2

1. kg/ha of Roundup's active ingredient (isopropylamine salt of glyphosate).

3.2 *Agrobacterium* Assay

Several strains of bacteria (all small gram-negative rods) were isolated from 11 of 25 subclone cultures sampled; the other shoots did not contain bacteria. When present in a shoot, bacteria were readily isolated from each of the four sections sampled. None of the isolated strains reacted positively (i.e. resulted in Cu_2O formation) in the Bernaerts-De Ley test (1), but *Agrobacterium* strain C58 always did.

4. DISCUSSION

Results of the present tests of NC-5339 and its subclones into which the altered *aroA* had been introduced indicated that the genetic transformation did not influence height growth or morphology in the absence of Roundup. Further, height growth data indicated that measurable but not total tolerance to Roundup was conferred by the altered *aroA* gene. Forestry application dosages for Roundup normally range from 0.5 to 1.5 kg/ha, but tolerance to these dosages was not observed in transformed, greenhouse-grown plants in the present tests.

The lack of total tolerance to Roundup observed in our tests may be attributable to several factors. First, the mannopine synthase promoter, which was fused to the altered *aroA* gene in this study, is known to be less effective in higher plants than other promoters such as the 35S promoter of cauliflower mosaic virus. Second, the native EPSP synthase is known to be localized primarily in chloroplasts (11) as are many other nuclear-encoded polypeptides (9). After translation, EPSP synthase moves from the cytoplasm to the chloroplast as an enzyme precursor that includes a transit peptide (14). The transit peptide is cleaved from the enzyme in the chloroplast, forming the active enzyme (4). The genetic construct used in the present transformation experiments lacked code for the transit peptide. Thus, the enzyme was only cytoplasmically active. Roundup tolerance greater than observed in the present tests might be achieved by using a more effective gene promoter or a transit peptide coding sequence in the *aroA* construct.

Results of our *Agrobacterium* tests indicated that shoot cultures of the NC-5339 subclones did not contain *Agrobacterium* strain C58/587/85, which had been used for their genetic transformation. However, some of the tested shoot cultures contained other types of bacteria. The *Agrobacterium* test we used has a precise basis: Some strains of *Agrobacterium*, including strain C58 (but not other bacteria), oxidize lactose to 3-ketolactose. This oxidation can be used as a test for *Agrobacterium* because, in the presence of Benedict reagent, 3-ketolactose forms readily visible, yellow Cu_2O on a blue background (1).

If necesary, bacterial contamination could be removed from poplar shoot cultures by two techniques. First, bacteria could be eliminated by repeated subculture of only very small shoot tips. Second, and more effectively, bacteria could be eliminated by rooting the shoots, growing them rapidly as greenhouse plants, and reestablishing *in vitro* shoot cultures from the uppermost auxillary nodes of the greenhouse plants. Rapid growth of the greenhouse plants was essential in the second method.

In summary, results from the present study indicated that Roundup tolerance in hybrid poplar NC-5339 was achieved by introducing an altered *aroA* gene using *Agrobacterium*-mediated genetic transformation. However tolerance, while detectable, was not at a commercial level; thus, improvements in the aroA construct, as described above, are required. Current results also demonstrated that transformed subclones of hybrid poplar NC-5339 can be freed of residual Agrobacterium contamination, which is important in vegetatively reproduced plants.

We thank Mr. M. White for research on the Agrobacterium assay, Mr. L. Petersen for technical assistance during spray testing, and Dr. L. Comai for supporting research, scientific advice, and technical review of the manuscript.

REFERENCES

1. Bernaerts, M. J. and J. De Ley. 1963. A biochemical test for Crown Gall bacteria. Nature 197:406-407.
2. Comai, L., D. Facciotti, W. R. Hiatt, G. Thompson, R. E. Rose, and D. M. Stalker. 1985. Expression in plants of a mutant aroA gene from Salmonella typhimurium confers tolerance to glyphosate. Nature 317:741-744.
3. Dandekar, A. M., P. K. Gupta, D. J. Durzan, and V. Knauf. 1987. Transformation and foreign gene expression in micropropagated Douglas-fir (Pseudotsuga menziesii). Bio/Technology 5:587-590.
4. della-Cioppa, G., S. C. Bauer, M. L. Taylor, D. E. Rochester, B. K. Klein, D. M. Shah, R. T. Fraley, and G. M. Kishore. 1987. Targeting a herbicide-resistant enzyme from Esherichia coli to chloroplasts of higher plants. Bio/Technology 5:579-584.
5. Fillatti, J. J., B. H. McCown, J. Sellmer, and B. Haissig. 1986. The introduction and expression of a gene conferring tolerance to the herbicide glyphosate in Populus NC 5339. In Proc. Tech. Assoc. of the Pulp and Paper Industry - 1986 Research and Development Conference, Raleigh, NC, TAPPI Press, Atlanta, GA. pp. 83-84.
6. Fillatti, J. J., J. Sellmer, B. McCown, B. Haissig, and L. Comai. 1987. Agrobacterium mediated transformation and regeneration of Populus. Mol Gen. Genet. 206:192-199.
7. Haissig, B. E. and D. E. Riemenschneider. 1987. Genetic engineering of hybrid poplars for herbicide tolerance. In Proc. Forest Products Research Conference 1986 - Matching Utilization Research with the Needs of Timber Managers, October 21-23, 1986. U.S. Dept. Agric., Forest Service, Forest Products Laboratory, Madison, WI, pp. 37-46.
8. Klee, H., R. Horsch, and S. Rogers. 1987. Agrobacterium-mediated plant transformation and its further applications to plant biology. Ann. Rev. Plant Physiol. 38:467-486.
9. Loffelhardt, W. 1987. Engineering chloroplasts: Prospects and limitations. Physiol. Plant. 69:735-741.
10. McCown, B. H. 1985. From gene manipulation to forest establishment: Shoot cultures of woody plants can be a central tool. TAPPI J. 68:116-119.
11. Mousdale, D. M. and J. R. Coggins. 1985. Subcellular localization of the common shikimate-pathway enzymes in Pisum sativum L. Planta 163:241-249.

12. Parsons, T. J., V. P. Sinkar, R. F. Stettler, E. W. Nester, and M. P. Gordon. 1986. Transformation of poplar by Agrobacterium tumefaciens. Bio/Technology 4:533-536.
13. Sederoff, R., A. Stomp, W. S. Chilton, and L. W. Moore. 1986. Gene transfer into loblolly pine by Agrobacterium tumefaciens. Bio/Technology 4:647-649.
14. Shah, D. M., R. B. Horsch, H. J. Klee, G. M. Kishore, J. A. Winter, N. E. Tumer, C. M. Hironaka, P. R. Sanders, C. S. Gasser, S. Aykent, N. R. Siegel, S. G. Rogers, and R. T. Fraley. 1986. Engineering herbicide tolerance in transgenic plants. Science 233:478-481.
15. Stalker, D. M., W. R. Hiatt, and L. Comai. 1985. A single amino acid substitution in the enzyme 5-enolpyruvylshikimate-3-phosphate synthase confers resistance to the herbicide glyphosate. J. Biol. Chem. 260:4724-4728.
16. Steinrucken, H. C. and N. Amrhein. 1980. The herbicide glyphosate is a potent inhibitor of 5-enolpyruvylshikimic acid-3-phosphate synthase. Biochem. Biophys. Res. Commun. 94:1207-1212.

TRANSFORMATION OF HYBRIDS *POPULUS TREMULA X P.ALBA* BY *AGROBACTERIUM TUMEFACIENS*.

M.F.MICHEL, F.DELMOTTE, and C.DEPIERREUX.

Station d'Amelioration des Arbres Forestiers, I.N.R.A.-C.R. d'Orleans, Ardon, 45160 Olivet, France

The improvement of *Populus* L. in France has been directed towards the selection of species and hybrids belonging to the Leuce Section (*Populus alba*, *P.tremula* and *P.tremuloïdes*). These hybrids are well suited to large scale planting in hydromorphic forest soils. However, these hybrids are highly susceptible to *Agrobacterium tumefaciens*, a pathogenic soil bacterium. The ability to induce galls and to regenerate plants from cells suggest that poplars may be an useful model for genetic engineering.

Two-week-old explants cultivated *in vitro* on half strength Murashige and Skoog medium were inoculated with two strains of *A.tumefaciens* (2516-2178) (1) isolated from poplars tumors.

We observed galls and teratoma formations on whole explants 8 weeks after inoculations.

Galls were transferred to medium without growth regulators but containing 1 mg/ml carbenicillin to kill the bacteria. We observed a fast growth of transformed cells and proliferation of shoots. Some of them showed a slow growth with thickened stem and were unable to root. The others were indistinguishable from untransformed control shoots and developped roots.

Nopaline was detecting in galls and in the two kinds of transformed shoots by paper electrophoresis at pH 3.9 stained with phenanthrenequinone. These results have to he confirmed by the Southern analysis. But yet they are according with those obtained by Fillatti (2) and De Verno (3).

REFERENCES

1. Nesme X., Michel M.F., and Digat B.,: Population heterogeneity of *Agrobacterium tumefaciens* in galls of *Populus* L. from a single nursery. Appl. Environ.Microbiol.53, 655-659, 1987.
2. Fillatti J.A.J., Sellmer J., MCCown B., Haissig B., and Comai L.,:*Agrobacterium* mediated transformation and regeneration of *Populus*. Mol.Gen. Genet.206, 192-199, 1987.
3. De Verno L.,: Variation in crown gall tumorigenesis among *Populus* clones. IUFRO,Molecular genetics working group. Petawawa, Forestry Institute. 16-18 june 1987.

M. R. Ahuja (ed.), Somatic Cell Genetics of Woody Plants, 81.

GENE TRANSFER IN WOODY PLANTS: PERSPECTIVES AND LIMITATIONS

M.R. AHUJA

Federal Research Centre of Forestry and Forest Products
Institute of Forest Genetics and Forest Tree Breeding
Sieker Landstrasse 2, D-2070 Grosshansdorf, Federal Republic of Germany

0. ABSTRACT

Conventional and more recently unconventional methods have been employed for transfer of genes in plants. Hybridization involving backcrosses has been effectively utilized in plant species for transfer of specific genes. However, this procedure takes too long, and would be impractical in long-lived tree species. By employing unconventional approaches involving *Agrobacterium* mediated gene transfer, or by direct gene transfer, transgenic plants have been produced in a number of plant species, including a tree species *Populus*. Dominant selectable markers (kanamycin resistance), herbicide tolerance and insecticidal genes, originally cloned from bacteria, have been transfered and expressed in plants. In some cases the transmission of the foreign genes to the progeny has been demonstrated. Unexpected genetic variability may be associated with foreign gene transfer in plants. This variability may be a cause of concern in long-lived tree species, following foreign gene transfer. Perspectives and limitations of foreign gene transfer will be discussed in plants, in particular tree species.

1. INTRODUCTION

Hybridization has played an important role in the transfer of genes from one species to another and the origin of cultivated plants. For example, wheat, cotton, potato, and tobacco are all polyploids (94) which probably originated by hybridization between their putative parents followed by chromosomal doubling/rearrangements in nature. Polyploidy is uncommon in forest tree species. Coastal redwood, *Sequoia sempervirens*, is considered a hexaploid having 2n=6x=66 chromosomes (41, 80, 88, 93), and may be a segmental allopolyploid or an autoallopolyploid (93). Naturally occuring triploids have been reported in European aspen, *Populus tremula*, (53) and quaking aspen, *Populus tremuloides*, (24). The triploid aspen (3n=57) found in the diploid (2n=38) stands were initially identified on the basis of leaf size and unusually long fibers, and later confirmed by chromosomes counts. Since triploid aspens have long fibers and greater volume, as compared to their diploid counterparts, there has been interest in polyploidy as an avenue for genetic improvement of aspen (39). Following crosses between tetraploid and diploid aspen, triploids have been produced both in Europe and the United States of America for testing their potential for biomass production. Since triploids are generally sterile, *in vitro* technology has been exploited for clonal propagation

M. R. Ahuja (ed.), Somatic Cell Genetics of Woody Plants, 83–101.

of triploid aspens (8, 10, 11, 20). Allopolyploids derived by hybridization between *Populus tremula* and *P. tremuloides* (64) have not been successfully exploited. Polyploidy associated with hybridization is one of the efficient ways of producing stable and fertile species from sterile interspecific hybrids. In such cases, there is transfer of complete sets of genes from two parents to the new genotype. Other methods of gene transfer involving one or a few genes are discussed in the subsequent sections of this review.

As it turns out, gene transfer has been described for many more herbaceous plants than for woody plants. Therefore, it is not possible to discuss general principles of gene transfer without extensively drawing on referees that deal with herbaceous species.

2. HYBRIDIZATION

2.1. Backcross

For improvement of plants involving specific genes, the breeder employs a backcrossing program. This is often combined with mutation selection for the introduction of new sets of genes. By employing backcrossing programs, it is possible to transfer one or a few genes from one species to another. Essentially this hybridization procedure, which may involve several generations, selects for a desirable gene(s) from one species on the genetic background of the second. By using this approach a number of useful genes have been transferred from a wild species *Aegilops umbellulata* to cultivated wheat, *Triticum aestivum* (81), and stem-rust resistance was transferred from *Agropyron elongatum* to common wheat (57). By employing interspecific hybridization, yellow mosaic resistance was transferred to cultivated mungbean, *Phaseolus aureus* from a wild relative *Phaseolus sublobatus* (18).

In addition to transfer of agronomically important genes into cultivated plants, genes have also been transferred for induction of genetic tumors in plants (apparently in the absence of any causative agent). Such systems provide insight into growth and differentiation, developmental timing of gene action, and they have relevance to foreign genes introduced into plants. Genetic tumors develop on plants of particular genotypes. These have been described in certain interspecific hybrids of several plant genera (3,91). The parental species involved in such tumorous hybrids are entirely non-tumorous. Genetic tumors have been investigated to a considerable extent in *Nicotiana* (2, 3, 4, 91), and *Lycopersicon* (16, 37, 63).

In the *Nicotiana* species hybrids, tumors generally appear during and after the flowering phase. However, stresses of different nature, such as wounding, chemical treatments, X-rays, crowding, etc., can initiate and enhance tumor formation at any stage and on any part of the hybrid plant (3). Whenever tumors are produced at the seedling stage, the plants do not survive. Apparently the sensitive system comes into existence in the hybrids between two unrelated species. Since the hybrid plants grow normally under optimal conditions until the flowering phase, it may be supposed that the action of genes responsible for tumor formation is suppressed at early stages of the hybrid plant development. The tumor genes are turned on at the time of flowering, which may involve a hormonal shift from the vegetative to the flowering or maturation phase (4). Tumors in Nicotiana exhibit a slight tendency to differentiate aberrant shoots, and thus resemble crown gall teratomas. Tumors themselves or tissues derived from tumorous plants are very difficult to root (5, 17). Tissues derived from tumor-prone plants exhibit autonomy with respect to phytohormones, such as auxin and kinetin (5, 17, 84).

The nature of genetic control of tumor formation has been investigated in Nicotiana. In one hybrid combination involving N. debneyi-tabacum (2n=96) and N. longiflora (2n=20), tumor formation was found to depend on the presence of a specific longiflora chromosome or its fragment, transferred by interspecific backcrossing, on the genetic background of debneyi-tabacum (2). Following genetic introgression, the longiflora chromosome fragment was no longer cytologically detectable, and presumably integrated in the debneyi-tabacum genome. Tumor condition in this hybrid combination was transmitted as a dominant trait (4). The main effect of the alien longiflora chromosome or its fragment (carrying the Tu or onc gene) on the debneyi-tabacum background seems to be on those loci involved in the biosynthesis of phytohormones (5, 17), that are presumably turned on continuously during the flowering phase of the tumor-prone hybrids in Nicotiana (4).

A genetically controlled tumor-like condition appears on a tomato hybrid system. The tumor condition develops on the segregants carrying the Frosty spot (Frs) gene from Lycopersicon chilense on the genetic background of L. esculentum variety Tiny Tim (63). Tumors, consisting of masses of enlarged parenchymatous cells, generally develop on the ventral surface of the third and subsequently formed leaves of the hybrid derivatives (37). These outgrowths do not differentiate into teratomas (c.f. genetic tumors in Nicotiana). Abortive floral buds develop on the tumorous plants, that are abscised after a short time. Plants with the tumor trait have a poorly developed vascular tissue, exhibit poor growth and are dwarfed, as compared to the normal segregants in the same population (16, 37). The above observations were based on the plants grown in the greenhouse.

Under field conditions, the picture is entirely different in the tomato tumor system. Tumor-prone plants grow vigorously in the field and lose all signs of tumors; flowers develop with little blossom drop, and fruits set and ripen, yielding viable seed. On the basis of morphology the tumor-prone and normal segregants are indistinguishable under field conditions (37). It would appear that Frs is a temperature/humidity sensitive gene, and high temperatures (above 28°C) and low humidity suppress the action of Frs gene on the genetic background of tomato.

Tissue culture studies have indicated that tissues from tumorous tomato plants do not exhibit autonomy with respect to growth hormones (37), that is generally associated with fully transformed tumors (22). Nevertheless, tumor and non-tumor tomato tissues are metabolically different (37). While the growth of tumor tissue was inhibited with increasing concentrations of auxin *in vitro*, the growth of the non-tumor tissues was enhanced by these concentrations (up to 10 mg/l). With respect to kinetin requirement, there were no significant differences between tumor and non-tumor tissues of tomato (37).

It would appear that the main effect of the dominant gene (Frs) from *Lycopersicon chilense* on the genetic background of *L. esculentum* is to raise slightly the endogenous levels of auxin in the tumor-prone genotypes. This alone, of course, does not cause growth autonomy *in vitro*. The tumor condition in tomato, controlled by a dominant gene, appears to be endowed with a trigger leading to tumor initiation, but very little stimulation for tumor development *in vivo* or *in vitro*. Perhaps this represents an early step in the direction of tumor formation (37).

Thus, clearly genes for specific traits have been transferred by hybridization in plants, whether for economic traits or for investigating developmental genetics. Although the process of gene transfer by this method is slow and time consuming, it is effective and transferred genes are generally stably transmitted through the seed. Because of long generation cycles, hybridization as an approach for single gene transfer in the forest tree species seems to have rather limited possibilities. But hybridization combined with recombinant DNA technologies could be explored for gene transfer in the forest trees.

3. FUSION OF PROTOPLASTS

Somatic cell hybridization offers alternatives to sexual hybridization, when the species are sexually incompatible. In plants protoplasts have been fused and in some combinations somatic hybrids have been produced. The genetic instability of somatic hybrids has been investigated in a number of cases. Partial chromosome losses, which may be random or preferential, have been observed in a number of somatic hybrids produced so far (44, 45, 50, 59, 67, 68, 70, 76). The unilateral loss of chromosomes of one parent in the somatic hybrid seemingly hold possibilities for gene transfer through protoplast fusion (90). In an intergeneric somatic hybrid involving *Aegopodium podograria* and *Daucus carota*, all chromosomes of *A. podograria* were apparently lost, as determined by chromosomal counts. However, molecular hybridization studies indicated that a small segment of *A. podograria* was integrated in the *Daucus carota* genome (38). In another intergeneric somatic hybrid produced by fusion of *Physalis minima* and *Datura innoxia* protoplasts, one to three chromosomes of *P. minima* were preferentially retained on the genetic background of *Datura innoxia* (46). In a recent study it was demonstrated that nucleolar chromosomes of *Solanum phureja* are eliminated preferentially in an interspecific somatic hybrid between *Solanum tuberosum* and *S. phureja* (76).

Inspite of the encouraging reports, protoplast fusion as an avenue for specific gene transfer has limited possibilities in plant species. This is because there is extensive variability and instability in the somatic hybrids, and most of the chromosomal losses are random in the somatic hybrids produced so far. The selection of desirable genotypes following fusion in a mixture of heterokaryons and homokaryons in culture may still present certain problems. The tree species pose a different kind of problem: they are recalcitrant and it is still very difficult to regenerate plants from the protoplasts of most forest tree species (7, 9, 12, 31, 66). In this direction, a recent report on the recovery of plants from the leaf protoplasts of *Populus* is encouraging (86).

Plant protoplasts have also been fused with animal cells (interkingdom fusions) to determine the fate of fusion products and expression of plant and animal genes (7). The range and extent of such experiments may provide clues to the coexistence of plant-animal genes in their hybrid, the so-called "Planimal" (7).

4. AGRIBACTERIUM PLASMID-VECTOR SYSTEM

4.1. Biology

4.1.1. *The Ti plasmid*. Genetic transformation of plants may be mediated by *Agrobacterium* carrying specific plasmids. One type of plasmid, the Ti plasmid, is carried by *A. tumefaciens* and induces tumors on a variety of different plant species. The second type, the Ri plasmid is present in *A. rhizogenes* and is involved in causation of hairy root disease in plants. The Ti plasmid of *A. tumefaciens* has been most extensively employed for transfer of genes in plants. Tumor formation is caused by transfer of T-DNA of the Ti plasmid in the nuclear genome of the plant cell, where it is subsequently stably integrated. The precise location where T-DNA is integrated in the genome remains unclear. Following integration, the T-DNA seems to be stably maintained in the cell lineage of the transformed cells (27, 61).

4.1.2. *The genes in T-DNA*. The Ti plasmid contains several genes for the initiation and maintenance of the tumorigenic state (71,85,87). Specifically, the T-DNA region of the Ti plasmid contains genes that are involved in the biosynthesis of phytohormones, auxins and cytokinins. Two genes (*iaaM* and *iaaH*) encode for the enzymes tryptophane monooxygenase (96) and indoleacetamide hydrolase (89, 95) and are involved in the synthesis of auxin, indoleacetic acid (IAA). At least one gene (ipt) encodes an enzyme isopentenyl transferase (19, 21), which is involved in the synthesis of the cytokinin, isopentenyl adenosine. Following transformation, the phytohormone genes from the T-DNA control the biosynthetic machinery of the plant cell and lead to the development of tumors. The transformed cells produce high levels of phytohormones necessary for the maintenance of the tumorous state (22).

In addition to the phytohormone genes (the so-called *onc* genes), the T-region also contains genes that code for opines, which are utilited by the bacteria for their growth. The opines are amino acid derivatives that are normally not metabolized by plant cells. On the basis of specific opines present, the crown gall tumors are classified as nopaline, octopine or agropine tumors.

4.1.3. *The virulence region*. The transfer of T-region from the Ti plasmid to the plant cell's genome requires the presence of the virulence (vir) region. The vir region is located in another part of the Ti plasmid. The activation of the vir region is mediated by specific phenolic compounds present in the exudates of wounded plant cells. One of the phenolic substances is acetosyringone (4-acetyl-2,6-dimethoxyphenol) (92).

4.1.4. The T-DNA borders. The ends or borders of T-DNA contain imperfect 25 base pair repeats (71, 85). Since the ends are the only region of the T-DNA that are essential for the transfer of the T-region to the plant cell, all genes originally present in the T-DNA can be removed without affecting the transfer. Any genetic sequences inserted between the direct repeats or ends of the T-DNA are co-transferred and integrated in the plant genome following transformation with *Agrobacterium* (85, 87). By employing disarmed Ti plasmids, gene transfer and recovery of normal plants have been demonstrated in a number of plant species (48, 51, 52, 62, 73).

4.2. Vectors for gene transfer

In the plant transformation mediated by *Agrobacterium*, two kinds of vectors systems have been employed: the cointegrating vectors and the binary vectors.

In the cointegrating vector system, the essential requirement is a region of genetic homology between the vector plasmid carrying the selectable marker gene and the Ti plasmid. The genetic homology between the vector and the Ti plasmid allows the integration of the vector into the Ti plasmid. Two types of such cointegrating vectors have been constructed by employing recombinant DNA technology. The first type consists of a disarmed *Agrobacterium* Ti plasmid, for example pGV3850 (101), in which the phytohormone genes are replaced by vector plasmid pBR322 sequences. In this construction, the cointegrating vector is placed between the T-DNA borders, along with the nopaline synthase gene from the Ti plasmid (49). A slightly different approach for constructing a cointegrating vector involved removal of the right border and phytohormone genes from the T-DNA. A vector plasmid with the right border, a selectable marker(s), and the region of homology is then cointegrated into the disarmed Ti plasmid containing the left border (42). The new cointegrating vector has both T-DNA borders and is functional as a gene transfer vector in the 'acceptor' plant species.

Binary vector systems are somewhat different than cointegrating vectors, as there is no region of genetic homology between the Ti plasmid and the vector plasmid. As a consequence there is no cointegration but mutual co-existence of the two types of plasmids in the *Agrobacterium*. The binary vectors contain a region for the origin of replication from a broad host-range plasmid (55, 58). The frequency of introducing vector plasmid into *Agrobacterium* for construction of binary vector is considerably higher (10^{-1}), as compared to cointegrating vector formation at 10^{-5} (56). Since the vector plasmid is not dependent on a specific Ti plasmid for a binary vector construct, the vectors carrying specific genes may be introduced virtually in any *Agrobacterium* strian containing a disarmed or an oncogenic Ti or Ri plasmid, as long as there is a vir region. As a result a wide range of binary vectors can be constructed for plant species where transformation may be limited by the host-range of the Agrobacterium strain.

Since the vector plasmids do not cointegrate, the binary vectors are not very stable and the vectors may be lost from Agrobacterium. The cointegrating system, on the other hand, is rather stable in Agrobacterium. There is evidence to support that a cointegrating vector can transform tomato at a higher frequency as compared to a binary vector (65).

4.2.1. Selectable marker genes. The oncogenic Agrobacterium tumefaciens contains selectable markers in the Ti plasmid that are descernible following transformation. These include: a) the ability of the transformed cells to grow on a hormone-free medium in vitro, whereas the untransformed cells require phytohormones for their growth, and b) the presence of opines in the transformed cells. Since oncogenic strains of A. tumefaciens induce tumors following transformation in vivo, which is an undesirable trait, other selectable markers had to be searched for an occurrence of a transformation event. The selectable marker system should be such that it allows a clear distinction between the transformed and untransformed cells: growth vs growth inhibition in vitro; presence or absence of an enzyme or a chemical. Some of the available selection systems employ chemical compounds, such as antibiotics, which specifically inhibit the growth of untransformed cells or gradually kill them.

One of the most widely used selectable markers has been the neomycin phosphotransferase, type II (NPT II) gene which confers resistance to the transformed cells against the antibiotic kanamycin. The enzyme NPT II detoxifies aminoglycoside compounds, such as kanamycin. The cells carrying the NPT II gene have a selective survival advantage in the presence of kanamycin, whereas the cells lacking the NPT II gene will be a selective disadvantage, resulting in their growth arrest and eventual death. However, kanamycin resistance gene does not seem to function in all plant species. There are so-called 'escapes', that is, untransformed cells may escape the kanamycin toxicity. This would imply that kanamycin may not be toxic in a given plant species, or the kanamycin resistance gene may not function as a selectable marker in the transformed cells. For this reason, other marker systems have been developed in which the selectable gene products confer resistance to compounds such as hygromycin, or methotrexate (56).

By employing recombinant DNA technology, chimeric genes have been constructed that utilize a broad spectrum promoter region of nopaline synthase (nos), a kanamycin resistance gene (NPT II), and a polyadenylation signal of octopine synthase (ocs) gene 48). Such chimeric genes are constructed and then inserted into disarmed Agrobacterium Ti plasmids. One such construct that we have employed in our laboratory is the A. tumefaciens strain C58CIpGV3850::1103, which carries a kanamycin resistance gene (48). In addition, we have employed the oncogenic nopaline (C58) and octopine (B6S3) strains of A. tumefaciens for the genetic transformation studies in the forest tree species. The Agrobacterium strains were kindly provided by Dr. J. Schell, Max Planck Institute, Cologne.

5. TRANSFORMATION IN TREE SPECIES

5.1. Tumor formation and growth autonomy

The woody plants are in general difficult to transform following infection with *Agrobacterium tumefaciens*. This may be due to host range of an *A. tumefaciens* strain, or that specific genotypes in a heterozygous population of a tree species are more porne to genetic transformation than others. Therefore, forest trees generally exhibit differential susceptibility to *A. tumefaciens* transformation. In an extensive investigation, among 112 forest trees tested, 65 species developed tumors of varying sizes and incidence following inoculation with *A. tumefaciens* (35). Some of the tree species that were resistant to *A. tumefaciens* transformation included *Fagus sylvatica*, and species of *Pinus*, *Picea*, and *Larix* (35). In recent years, by employing different strains of *A. tumefaciens*, variable tumor formation has been reported in *Pinus teada* (82, 83), *Pinus sylvestris* (14), *Picea abies* (14, 28), *Picea nordmanniana* (28), *Pseudotsuga menziesii* (33), *Betula verrucosa* and *B. mandschurica* (99), *Populus tremula* and *P. tremula* x *P. tremuloides* (14).

Tissue culture studies have demonstrated growth autonomy of transformed cells, following inoculation with oncogenic *A. tumefaciens*, in *Pinus taeda* (82), *Pseudotsuga menziesii* (33), and *Populus tremula* (14).

5.2. Gene transfer with selectable markers

Dominant selectable marker (kanamycin resistance) has been employed in several woody plants to monitor gene transfer mediated by *Agrobacterium*. Transformation and expression of a foreign gene confering kanamycin resistance has been demonstrated in the micropropagated plants of *Populus* (40), and *Pseudotsuga menziesii* (33). *In vitro* selection for kanamycin resistant cells following transformation with disarmed *Agrobacterium* has been carried out in apple (32), and larch, *Larix* (54).

We have carried out transformation experiments in aspen (*Populus tremula*) and hybrid aspen (*Populus tremula* x *P. tremuloides*), birch (Betula pendula) and coastal redwood (*Sequoia sempervirens*) by employing the disarmed *A. tumefaciens* strain C58CIpGV3850::1103, carrying the kanamycin resistant gene (14). Although no transformants were recovered in the redwood, transformed microshoots from aspen and birch survived on media containing 100 mg/l to 200 mg/l kanamycin (14). Rooting of microshoots has been difficult on the media containing 60 mg/l to 100 mg/l kanamycin. In addition to expression of kanamycin resistance gene in the forest tree species, a commercially important herbicide tolerant gene, *aroA* gene from *Salmonella*, has also been transferred to *Populus* (40). However, there was a weak expression of the herbicide tolerant gene in the transgenic *Populus* plants (79). Efforts are now being made to attach a strong promotor to the *aroA* gene for enhanced expression of this gene (79).

6. TRANSFER OF FOREIGN GENES WITHOUT AGROBACTRIUM MEDIATION

6.1. The delivery systems

Gene transfer has also been accomplished without *Agro-*

bacterium mediation. For direct gene transfer into protoplasts one requires totipotent protoplasts and a delivery system - a vector system - that is protected against endonucleases present in the cell. In one system, plasmid DNA was fused with bacterial spheroplast and then transferred to Nicotiana protoplasts (47). Another delivery system involved the use of liposomes, which are artificial lipid vehicles. The DNA-liposome complex is highly tolerant to digestion by endonucleases. The fate of liposome-encapsulated DNA has been described in plant protoplasts (25, 72).

6.2. Direct gene transfer into protoplasts

In recent years direct gene transfer has been described in several plant species. By employing selectable dominant marker gene (kanamycin resistance), calf thymus carrier DNA, and totipotent protoplasts, stable transformation and transgenic plants have been regenerated in Nicotiana (77). Direct gene transfer has also been described in the protoplasts of monocots, as Lolium multiflorum (78), Triticum monococcum (60), and rice, Oryza sativa, (97). However, in these cases the protoplasts could only be carried upto the callus stage. Since monocots are not readily transformed by Agrobacterium, the method of direct gene transfer has potential for producing transgenic plants in this group of plants. However, the protoplasts of monocots in general are recalcitrant, and therefore direct gene transfer through protoplasts may have limited possibilities. Recent reports on regeneration of plants from rice protoplasts (1, 100) is certainly a positive development in this direction (29).

6.3. Direct injection of DNA

6.3.1. Microinjection. The prospects of foreign gene transfer by microinjection of DNA into protoplasts and cells have been discussed (30). In a recent study, random portions of plasmid DNA were found to be integrated into tobacco protoplasts following microinjection (30). This method of gene transfer would have limited possibilities in plant species with recalcitrant protoplasts, such as tree species. On the other hand, if microinjection could become an efficient method of gene transfer in cells or tissues, then this method might offer far reaching possibilities in plant species, in particular woody plants where protoplasts are still difficult to manipulate.

6.3.2. Injection into plants. In a recent study DNA has been directly injected into rye, Secale cereale, plants and expressed in the progeny (36). Plasmid DNA carrying the kanamycin resistant gene was injected into young male floral shoots of rye plants. Following cross pollination with the 'transformed pollen', a small proportion of the progeny (7 of 3000 seedlings) exhibited the kanamycin resistant trait. The efficiency of genetic transfer seems to be rather low with direct injection of DNA into plants. Several factors might influence the outcome. These include the genotype of the plant, number of available competent cells in the target area, the copy number of the injected gene, stability of the foreign gene in the host genome, and DNA rearrangements caused by the foreign gene.

7. GENETICS OF THE TRANSGENIC PLANTS

One of the key questions regarding genetic transformation relates to the stability of a foreign gene in the transgenic plants and their progeny. Earlier studies (23) indicated that the T-DNA carrying the tumor-inducing genes from the Ti plasmid was not very likely transmitted through meiosis to the next generation. However, later investigations with the deletions of phytohormone genes have revealed that T-DNA was stably transmitted through both male and female gametes, and the marker gene for octopine synthase (ocs) on the T-DNA was inherited as a dominant trait (74). This could be interpreted to mean that partial or complete removal of onc-genes, which result in apparently normal differentiation of shoots and roots, allows the T-DNA to move across the meiosis barrier; however, intact T-DNA probably does not. In the *Nicotiana* genetic tumor system, where teratomatous and fertile plants were recovered, the tumor condition was transmitted as a dominant trait (4). In this respect, it may be conjectured that genetic tumors in *Nicotiana* may be comparable to teratomatous crown galls tumors in their genetic constitution of the onc-genes, although genetic homologies at the molecular level have not been investigated in the two systems so far.

7.1. Inheritance of selectable markers

Subsequent studies on the transfer of T-DNA carrying foreign genes, or foreign DNA transferred without the aid of *Agrobacterium*, have revealed that genetics of transformants is somewhat more complicated than was anticipated. Previously a number of investigations have indicated that selectable marker (kanamycin resistance) is dominantly inherited in the sexual crosses of *Nicotiana tabacum* (48, 52, 77), *Nicotiana plumbaginifolia* (51), *Petunia* (52), and *Lycopersicon esculentum* (52, 65). However, in a number of cases there were exceptions. Either the tissue escaped the selection pressure of kanamycin ('escapes') or the kanamycin resistant gene was lost or was not expressed in the plants (52, 65, 77). In one study (52) 25 microshoots isolated from *Petunia* leaf discs, following genetic transformation for kanamycin resistance, were rooted in the presence of kanamycin. Of the 25 plants, 20 were kanamycin resistant, including five that showed delayed resistance to kanamycin. There were five plants that were sensitive to kanamycin. In the kanamycin sensitive plants either the kanamycin resistant gene was lost or its expression turned off (52) during development. Similar results were also reported in tomato (65). Nopaline expression was found in 12 of 25 plants in soil, and 17 of 25 in callus cultures (52). In a selfed progeny of a cultured-induced nopaline synthase, only two of 24 *Petunia* seedlings showed faint traces of nopaline; however, when the same seedlings were transferred to culture, 17 of 24 exhibited strong nopaline production (52).

Transgenic plants have been produced in *Nicotiana tabacum* that carry the toxic gene bt2 cloned from *Bacillus thurigiensis* (98). The protein produced by the gene bt2 is highly toxic to a variety of insect species. In these studies disarmed *Agrobacterium* strains carrying 4 chimeric genes containing insect toxin gene *bt2* (one with complete *bt2* coding sequence and three truncated *bt2* genes) along with a chimeric kanamycin resistance gene were employed for genetic transformation of tobacco leaf discs (98). A total of 85 transgenic tobacco plants, transformed by 4 types of chimeric *bt2* genes expressed various levels of active insect toxin, when tested on the lepidopteran larvae of *Manducta sexta*, a tobacco pest. The transgenic plants grew normally under green house conditions. From these 11 plants were tested for kanamycin resistance, and all contained at least one copy of the kanamycin resistance gene. The F-1 progeny of two plants showed large amounts of toxin protein (Bt2) and exhibited kanamycin resistance in all plants, and there were at least five copies of the T-DNA in these plants. This indicates that insect toxicity is correlated with the kanamycin resistance marker in the F-1 progeny. The expression of the complete *bt2* toxin gene and its truncated forms were quite different in the transformed plants. The truncated *bt2* gene sequences exhibited much higher levels of insect toxin as compared to the complete *bt2* gene, even though all these constructs were under the control of the same promotor. Why the complete *bt2* gene is expressed poorly in the transgenic plants is not known. A small proportion of F-1 progeny of two transgenic plants, which were produced following transformation with the truncated *bt2* gene and showing high toxin levels, were analysed. All 15 plants tested showed 100 percent killing of the insect, *M. sexta*, larvae in a standard test. The toxin levels in the F-1 progeny varied between 20 and 50 ng per mg of total protein, as compared to 30 ng per mg in the two parental transgenic plants (98).

Transgenic plants have also been produced in a tree species, involving a *Populus* hybrid, *P. alba* x *P. grandidentata*, (40). Leaf explants were transformed with: a) an oncogenic *Agrobacterium tumefaciens* nopaline strain C58 harboring a binary vector carrying three chimeric genes, of which two are for kanamycin resistance and the third *aro*A gene for herbicide (glyphosate) tolerance, and b) the same construction as above but in a disarmed (non-oncogenic) strain of *A. tumefaciens*. The selection for the transformation event was carried out on a kanamycin (60 mg/l) medium. Microshoots were not recovered on a selection medium following co-cultivation of leaf explants with the oncogenic *Agrobacterium* containing the binary vector. However, two types of microshoots developed on the leaf explants following transformation with the oncogenic strian carrying the binary vector. The morphology of type A shoots was normal, while that of type B shoots was abnormal and teratomatous. In the type A shoots, 38 of 40 showed NPT II activity (kanamycin resistance), but none of the 40 shoots exhibited nopaline synthase activity. On the other hand, in the type B shoots, 18 of 20 showed the NPT II activity and all 20 were positive for nopaline synthase activity (40). In a different *Populus* hybrid (*P. trichocarpa* x *P. deltoides*) transformation was obtained in cultured cells, but there was no recovery of transgenic plants (75).

7.2. Somatoclonal Variation

Thus it would appear that genetic transformation leading to transgenic plants depends upon the genotype of the host and the *Agrobacterium* strain or the vector system emplyed. A foreign gene may or may not be expressed in the transformed plant or its progeny. Variability recovered in the somatic tissues *in vitro* that seems to be associated with genetic transformation and gene transfer events may be termed "Somatoclonal Variation" (14), as compared to the general term somaclonal variation. The letters *to* inserted in the somaclonal indicate the possible effect of a genetic transformation/ foreign gene transfer event. Somatoclonal variation may be due to: 1) loss of a foreign gene, 2) loss of expression of a foreign gene, 3) rearrangements caused by a foreign gene, 4) position effect, 5) copy number of a foreign gene inserted into the host genome (one chromosome or different chromosomes), and 6) the location where the foreign gene is inserted in the genome (euchromatin or heterochromatin).

8. PERSPECTIVES AND LIMITATIONS

A number of different approaches have been employed for the transfer of foreign genes into different plant species. Hybridization still remains an effective force in this domain, as this will be always required for checking the transmission/ inheritance of a foreign gene. Foreign genes must be properly transported before they can enter the plant cell doors leading to the master molecule (DNA). Most of the transportation designs in this sophisticated area have been innovated by the recombinant DNA technology. Not only that the foreign genes must have transportation chariots, but also they must bring a password to enter the cell's genome. That password is the promoter region for recognition of transcriptional signal in the plant cell. In addition, foreign genes must bring a termination signal, that they would require at the end of a hard days work! This is the beginning of a new area. Genes can be isolated, sequenced, cloned, put in transporting vehicles (vectors) and transferred at will from one living organism to another. However, there are limitations.

Foreign genes have been introduced and expressed in a number of plant species. Transgenic plants have been produced in *Nicotiana tabacum* (34, 52, 77, 98), *Nicotiana plumbaginifolia* (51), *Lycopersicon esculentum* (65), *Arabidopsis* thaliana (62), and *Brassica napus* (43).

Among the tree species, transgenic plants have been produced in *Populus* (40). The herbicide tolerant gene (*aro*A) has been transferred and expressed in transgenic *Populus* plants. Transfer of other desirable foreign genes may follow. But do we have other economically important genes at our disposable that could be transferred to the tree species? Most of the economically important traits, such as growth, vigor, yield, height and biomass are quantitative traits, controlled by polygenes. These are rather complex and poorly understood at the molecular level. Since recombinant DNA technology offers better prospects for single gene or a block of closely linked gene transfer, we would be limited to single or a few gene transfer at the present time. There are some other traits that might be considered for gene transfer are drought tolerance, salt tolerance, and frost tolerance.

Resistance to diseases and pests seems to be controlled by major genes. Insect toxic gene bt2 from *Baccilus thuringiensis*, already transferred in tobacco, could be a potential foreign gene for transfer in tree species. Another candidate might be the blister rust resistance gene in pines. Blister rust resistance gene has been detected in sugar pine, *Pinus lambertiana*. But the problem is how to identify, isolate and clone this gene from sugar pine. Very little information is available regarding the structure of the genome, and linkage maps are virtually non-existent in the forest tree species. Therefore, there is urgent need for research in cytogenetics and molecular genetics of forest tree species.

In the initial stages, genetic engineering in forestry will rely on foreign genes obtained from widely unrelated organisms, for example, bacteria, herbaceous plant species, or even animals. Although vailability of *in vitro* regenerating systems seems essential for transfer of foreign genes at the present time, recent experiments with direct injection of DNA into plants (36) might suggest a different approach. For example, plasmid DNA with a desirable dominant selectable marker could be injected into young male flowers/cones before the onset of meiosis. The pollen from such injected plants could then be used for pollination of normal female flowers/cones. The transgenic plants, if any, could be selected in the seedlings on the basis of the dominant selectable marker in the first seed generation, without the aid of tissue culture.

Because of certain limitations in the gene technology, we should continue to search for new variants by other methods in the forest tree species. But then we might ask the following questions. Is it really necessary to introduce foreign genes in the tree species? Have we already effectively utilized the enormous existing variability in the forest tree species? Have we adequately explored the potential of somaclonal variation/mutation selection (13) in the forest tree genetics and improvement? We do not have satisfactory answers to these questions at the present time. In the meanwhile, new variants may be searched by employing *in vitro* technology. By challenging cells to herbicides, plants tolerant to herbicides have been isolated in tobacco (26) and *Populus* (69).

The application of genetic engineering to forest trees raises certain genuine fears and concerns. Forest trees have long generation cycles, with an extended vegetative phase, ranging from ten to hundred years or more. Foreign genes may be immediately expressed, or remain inactive for a long time, or they might be lost during the long vegetative phase of the tree species. The foreign genes might cause genetic changes by position effects or rearrangements immediately or after long periods of time. Some of the genetic changes might be detected during the vegetative phase, while others might be detected during the flowering phase, that is, after maturity. It is possible that certain genetic changes (6) may 'turn on' the cellular 'tumor genes' or 'oncogenes' presumably present in the genomes of multicellular organisms (15), including forest trees. The oncogenes might be expressed before the flowering phase or after it, as in the genetic tumor system in *Nicotiana* (3,4) and lead to the development of tumors or cause growth abnormalities. These abnormalities might occur some ten to 50 years after the initial foreign gene transfer event. At least at the present time it is hard to predict how stable or unstable a foreign gene(s) would be in a long-lived tree species. Therefore, we should move very cautiously in the area of foreign gene transfer in the tree species.

REFERENCES

1. Abdullah C, EC Cocking and JA Thompson: Efficient plant regeneration from rice protoplasts through somatic embryogenesis. Bio/Technology 4: 1087-1090, 1986.
2. Ahuja MR: A Cytogenetic study of heritable tumors in *Nicotiana* species hybrids. Genetics 47: 865-880, 1962.
3. Ahuja MR: Genetic control of tumor formation in higher plants. Quart. Rev. Biol. 40: 329-340, 1965.
4. Ahuja MR: An hypothesis and evidence concerning the genetic components controlling tumor formation in *Nicotiana*. Mol. Gen. Genet. 103: 176-184, 1968.
5. Ahuja MR: Genetic Control of Phytohormones in tumor and non-tumor genotypes in *Nicotiana*. Ind. J. Exptl. Biol. 9: 60-68, 1971.
6. Ahuja MR: On the nature of genetic change as an underlying cause for the origin of neoplasms. In Antivival Mechanisms in the Control of Neoplasia, P Chandra (ed), Plenum Press, New York, pp. 17-37, 1979.
7. Ahuja MR: Isolation, Culture and fusion of protoplasts: problems and prospects. Silvae Genet. 31: 66-77, 1982.
8. Ahuja MR: Somatic cell genetics and rapid clonal propagation of aspen. Silvae Genet. 32: 131-135, 1983.
9. Ahuja MR: Protoplast research in woody plants. Silvae Genet. 33: 32-37, 1984.
10. Ahuja MR: A commercially feasible micropropagation method for aspen. Silvae Genet. 33: 174-176, 1984.
11. Ahuja MR: Aspen. In Handbook of Plant Cell Culture, DA Evans, WR Sharp and PJ Ammirato (eds), Macmillan Publishing Company, New York, pp. 626-651, 1986.
12. Ahuja MR: Perspectives in plant biotechnology. Curr. Sci. 55: 217-224, 1986.
13. Ahuja MR: Somaclonal variation. In Cell and Tissue Culture in Forestry, vol. 1, JM Bonga and DJ Durzan (eds), Martinus Nijhoff Publishers, Dordrecht, pp. 272-285, 1987.
14. Ahuja MR: Gene transfer in forest trees. In: Genetic Manipulation of Woody Plants, J Hanover and D Kiethley (eds), Plenum Press, New York (in press), 1987.
15. Ahuja MR and F Anders: Cancer as a problem of gene regulation. In Recent Advances in Cancer Research: Cell Biology, Molecular Biology, and Tumor Virology, RC Gallo (ed), CRC Press, Cleveland, Ohio pp. 103-117, 1977.
16. Ahuja MR and GR Doering: Effect of Gibberellic acid on genetically controlled tumor formation and vacularization in tomato. Nature 216: 800-801, 1967.

17. Ahuja MR and GL Hagen: Morphogenesis in Nicotiana debneyi-tabacum, N. longiflora and their tumor-forming hybrid derivatives in vitro. Dev. Biol. 13: 408-423, 1966.
18. Ahuja MR and BV Singh: Induced genetic variability in mumg-bean through interspecific hybridization. Indian J. Genet. & Pl. Breed. 37: 13-136, 1977.
19. Akiyoshi DE, et al.: T-DNA of Agrobacterium tumefaciens encodes an enzyme of cytokinin biosynthesis. Proc. Natl. Acad. Sci. USA 81: 5994-5998, 1984.
20. Barocka KH, M Baus, E Lontke and F Sievert: Tissue Culture as a tool for in vitro mass propagation of aspen. Z. Pfanzenzüchtg. 94: 340-343, 1985.
21. Barry GF et al.: Identification of cytokinin genes and transfer into plants. Curr. Top. Plant Biochem. Physiol. 4: 101-109, 1985.
22. Braun AC: A physiological basis of antonomous growth of crown gall tumor cell. Proc Natl Acad Sci USA 44: 344-349, 1958.
23. Braun AC and HN Wood: Supression of the neo-plastic state with the acquisition of specialized functions in cells, tissues and organs of crown gall teratomas of tobacco. Proc Natl Acad Sci USA 73: 496-500, 1976.
24. Buijtenen JP van, PN Joranson and DW Einspahr: Naturally occuring triploid quaking aspen in the United States. Proc. Soc. Am. For.: 62-64, 1957.
25. Caboche M and A Deshayes: Utilization de loposome pour la transformation de protoplasts de mesophylle de tabac par plasmide recombinant de E. Coli leur conferant la resistance à la Kanamycine. C.R. Acad. Sci. (Paris), 229: 663-666, 1984.
26. Chaleleff RS and TB Ray: Herbicide resistant mutants from tobacco cell cultures. Science 223: 1148-1151, 1984.
27. Chilton MD et al.: Stable incorporation of plasmid DNA into higher plants: The molecular basis of crown gall tumorigenesis. Cell 11: 263-271, 1977.
28. Clapham DH and I Ekberg: Induction of tumors by various stains of Agrobacterium tumefaciens on Abies nordmanniana and Picea abies. Scand. J. For. Res. 1: 435-437, 1986.
29. Cocking EC and MR Davey: Gene transfer in cereals. Science 236: 1259-1262, 1987.
30. Crossway A et al: Micromanipulation techniques in plant manipulation. Biotechniques 4: 320-334, 1986.
31. David A: Conifer protoplasts. In Cell and Tissue Culture in Forestry, vol. 2, J.M. Bonga and D.J. Durzan (eds.), Martinus Nijhoff Publishers, Dordrecht, pp. 2-15, 1987.
32. David DJ, AJ Passey, S Fredieri and E. Rugini: Regeneration and transformation of apple plants using wild type and engineered plasmid in Agrobacterium spp. This volume.
33. Dandekar AM, Gupta, PK, Durzan DJ and Knanf V: Transformation and foreign gene expression in micropropagated Douglas-fir (Pseudotsuga menziesii). Bio/Technology 5: 587-590, 1987.
34. De Block M et al: Expression of foreign genes in regenerated plants and in their progeny. EMBO J. 3: 1681-1689, 1984.

35. De Cleene M and L De Ley: The host range of crown gall. Bot. Rev. 42: 389-466, 1976.
36. De la Pena A, H Lörz and J Schell: Transgenic rye plants obtained by injecting DNA into young floral tillers. Nature 325: 174-176, 1987.
37. Doering GR and MR Ahuja: Morphogenetic studies of a genetically controlled tumor-like condition in Lycopersicon hybrids. Planta (Ber) 75: 85-93, 1967.
38. Dudits D et al.: Plant regeneration from intergeneric cell hybrids. Plant Sci.Lett.15: 101-112, 1979.
39. Einspahr DW and LL Winton: Genetics of quaking aspen. Forest Service, US Department of Agriculture, US Govt. Printing Office, Washington, pp. 1-23., 1977.
40. Fillatti JJ et al.: *Agrobacterium* mediated transformation and regeneration of *Populus*. Mol.Gen.Genet. 206: 192-199, 1987.
41. Fozdar BS and WJ Libby: Chromosomes of *Sequoia sempervirens*: 8-hydroxyquinoline-caster oil pretreatment for improving preparation. Stain Tech. 43: 97-100, 1968.
42. Fraley RT et al.: Expression of bacterial genes in plant cells. Proc. Natl. Acad. Sci. USA 80: 4803-4807, 1983.
43. Fry JE and RB Horsch: Transformation of *Brassica napus* by *Agrobacterium tumefaciens* based vectors. Proc. 6th Int. Congr. Plant Tissue Cell Culture (Abstracts), p. 127, 1986.
44. Gleba YY and F Hoffmann: *Arabidobrassica*: A novel plant obtained by protoplast fusion. Planta (Berl) 149: 112-117, 1980.
45. Goodman RM, H Hauptli, A Crossway and V Knauf: Gene transfer in crop improvement. Science 236: 48-54, 1986.
46. Gupta PP, O Schieder and M Gupta: Intergeneric nuclear gene transfer between somatically and sexually imcompatible plants through asymmetric protoplast fusion. MolGen.Gen.197: 30-35, 1984.
47. Hain R, HH Steinbiss and J Schell: Fusion of *Agrobacterium* and *E. Coli* spheroplasts with *Nicotiana tabacum* protoplasts - Direct gene transfer from microorganisms to higher plants. Plant Cell Rep. 3: 60-64, 1984.
48. Hain R et al.: Uptake, integration, expression and genetic transmission of a selectable chimaeric gene by plant protoplasts. Mol. Gen. Genet. 199: 161-168, 1985.
49. Herrera-Estrella L et al.: *Agrobacterium* as a vector for the introduction of genes into plants. In: Plant Genetic Engineering, JH Dodds (ed.), Cambridge University Press, Cambridge, pp. 61-93, 1985.
50. Hoffmann F and Adach T: *Arabidobrassica*: Chromosomal recombination and morphogenesis in asymmetric intergeneric hybrid cells. Planta (Berl.) 153: 586-593, 1981.
51. Horsch RB et al.: Inheritance of functional foreign genes in plants. Science 223: 496-498, 1984.
52. Horsch RB et al.: A simple and general method for transferring genes into plants. Science 227: 1229-1281, 1985.
53. Johnsson H: Cytological studies of diploid and triploid *Populus tremula* and of crosses between them. Hereditas 26: 321-352, 1940.

54. Karnosky DF, AM Diner and WM Barnes: A model for gene transfer in conifers: European larch and Agrobacterium. This volume.
55. Klee HJ, MF Yanofsky and EW Nester: Vectors for transformation of higher plants. Bio/Technology 3: 637-642, 1985.
56. Klee, H, R. Horsch and S Rogers: Agrobacterium-mediated plant transformation and its further applications to plant biology. Ann. Rev. Plant Physiol. 38: 467-486, 1987.
57. Knott DR: The inheritance of rust resistance. IV. The transfer of stem rust resistance from Agropyron elongatum to common wheat. Can. J. Pl. Sci. 41: 109-123, 1961.
58. Koncz C and J Schell: The promotors of T-L-DNA gene 5 controls the tissue-specific expression of chimerics carried by anovel type of Agrobacterium binary vector. Mol. Gen. Genet. 204: 383-396, 1986.
59. Krumbiegel G and O Schieder: Selection of Somatic hybrids after fusion of protoplasts from Datura innoxia Mill and Atropa belladonna L. Planta (Berl) 145: 371-375, 1979.
60. Lörz H, B Baker and J Schell: Gene transfer to cereal cells mediated by protoplast transformation. Mol Gen. Genet. 199: 178-182, 1985.
61. Lemmers M et al.: Genetic identifications of functions of TL-DNA transenpts in octopine crown gall. EMBO J. 1: 147-152, 1980.
62. Lloyd AM et al.: Transformation of Arabidopsis thaliana with Agrobacterium tumefaciens. Science 234: 464-466, 1986.
63. Martin FW: Frosty spot. A developmental disturbance of tomato leaf. Ann. Bot. 30: 701-709, 1966.
64. Mattila RE: On the production of tetraploid hybrid aspen by colchicine treatment. Hereditas 47: 631-640, 1961.
65. McCormick S., J Niedermeyer, J Fry, A Barnason, R Horsch and R Fraley: Leaf disc transformation of cultivated tomato (Lycopersicon esculentum) using Agrobacterium tumefaciens. Plant Cell Rep. 5: 81-84, 1986.
66. McCown BH and JA Russel: Protoplast culture of hard woods. In Cell and Tissue Culture in Forestry, vol. 2, JM Bonga and DJ Durzan (eds), Martinus Nijhoff Publishers, Dordrecht, pp. 16-30, 1987.
67. Melchers G: Protoplast fusion, mechanisms and consequences for potato breeding and production of potatoes + tomatoes. In Advances in Protoplast Research, L. Ferency and G.L. Farkas (eds), Pergamon Press, Oxford, pp. 283-286, 1980.
68. Melchers G, MD Sacristan and AA Holder: Somatic hybrid plants of potato and tomato regenerated from fused protoplasts. Carlsberg Res. Commun. 43: 203-218, 1978.
69. Michler CH and BE Haissig: Somaclonal selection of hybrid poplars that tolerate herbicides. This volume.
70. Nehls R et al.: Development of protoplast fusion products. In: Differentiation of Protoplasts and of Transformed Cells J Reinert and H Binding (eds), Springer Verlag, Berlin, pp. 67-108, 1986.
71. Nester EW, MP Gordon, RM Amasino and MF Yanofsky: Crown gall: a molecular and physiological analysis. Ann. Rev. Plant Physiol. 35: 387-413, 1984.

72. Ohgawara T, H Uchimiya and H Harada: Uptake of liposome-encapsulated plasmid DNA by plant protoplasts and molecular fate of foreign DNA. Protoplasma 116: 145-148, 1983.
73. Ooms G, MM Burrell, A Karp and J Hille: Genetic transformation in two potato cultivars with T-DNA from disarmed *Agrobacterium*. Theor. Appl. Genet. 73: 744-750, 1987.
74. Otten L, et al: Mendelian transmission of genes introduced into plants by the Ti-plasmid of *Agrobacterium tumefaciens*. Mol. Gen. Genet. 183: 209-213, 1981.
75. Parson TJ, VP Sinkar, RF Stettler, EW Nester and MP Gordon: Transformation of poplar by *Agrobacterium tumefaciens*. Bio/Technology 4: 533-536, 1986.
76. Pijnakar LP, MA Ferwerda, KJ Puite and S Roest: Elimination of *Solanum phureja* nucleolar chromosomes in *S. tuberosum* + *S. phureja* somatic hybrids. Theor. Appl. Genet. 73: 878-882, 1987.
77. Potrykus I, RD Shillito, M Saul and J Paszkowski: Direct gene transfer-state of the art and future potential. Plant Molec. Biol. Rep. 3: 117-128, 1985.
78. Potrykus I, M Saul, I Petruska, J. Paszkowski and RD Schillito: Direct gene transfer to cells of monocots. Mol. Gen. Genet. 199: 183-188, 1985.
79. Riemenschneider DE ae al.: Expression of an herbicide tolerance gene in young plants of a transgenic hybrid poplar clone. This volume.
80. Saylor LC and HA Simons: Karyology of Sequoia sempervirens: karyotype and accessory chromosomes. Cytologia 35: 294-303,
81. Sears ER: The transfer of leaf-rust resistance from *Aegilops umbellulata* to wheat. Brookhaven Symp. Biol. 9: 1-22, 1956.
82. Sederoff R, A Stomp, WC Chilton and LW Moore: Gene transfer into loblally pine by *Agrovacterium tumefaciens*. Bio/Technology 4: 647-649, 1986.
83. Sederoff R, et al: Application of recombinant DNA techniques to pines: a molecular approach to genetic engineering in forestry. In Cell and Tissue Culture in Forestry, vol. 1, JM Bonga and DJ Durzan (eds), Martinus Nijhoff Publishers, Dordrecht, pp. 314-329, 1987.
84. Schaeffer GW and HH Smith: Auxin-Kinetin interaction in tissue culture of *Nicotiana* species and tumor conditioned hybrids. Plant Physiol 38: 291-297, 1983.
85. Schell J, et al: Transfer of foreign genes to plants and its use to study developmental processes. In Cell Fusion, Gene Transfer and Transformation, RF Beers and EG Bassett (eds), Raven Press, New York, pp. 113-128, 1984.
86. Russel JA and BH McCown: Culture and regeneration of *Populus* leaf protoplasts isolated from non-seedling tissue. Plant Sci. 46: 133-142, 1986.
87. Schilperoort RA: Integration, expression and stable transmission through seeds of foreigns in plants. In Genetic Manipulation in Plant Breeding, W Horn, CJ Jensen, W Odenbach and O Schieder (eds), Walter de Gruyter and Co., Berlin, pp. 837-858, 1986.
88. Schlarbaum SE and T Tsuchiya: A chromosome study of coast redwood, *Sequoia sempervirens* (D.Don. Endl.). Silvae Genet. 33: 56-62, 1984.

89. Schroder G et al.: The T-region of Ti plasmids codes for an enzyme synthesizing indole-3-acetic acid. Euro. J. Biochem. 138: 387-391, 1983.
90. Shepard JF et Al.: Genetic transfer in plants through interspecific protoplasts fusion. Science 219: 683-688, 1983.
91. Smith HH: Plant genetic tumors. Progr. Exptl. Tumor Res. 15: 138-164, 1972.
92. Stachel SC, B Timmermann and P Zambryski: Generation of single standed T-DNA molecules during the initial stages of T-DNA transfer from Agrobacterium tumefaciens. Nature 322: 706-712, 1985.
93. Stebbins GL: The Chromosomes and relationship of Metasequoia and Sequoia. Science 108: 95-98, 1948.
94. Stebbins GL: Variation and Evolution in Plants. Columbia Univ. Press, New York, 1950.
95. Thomashow LS, Reeves S and Thomashow MF: Crown gall oncogenesis: evidence that a T-DNA gene from the Agrobacterium Ti plasmid pTiAG encodes an enzyme that catalyzes synthesis of indoleacetic acid. Proc. Natl. Acad. Scie. USA, 81: 5071-5075, 1984.
96. Thomashow MF, Hugly S, Buchholz W and Thomashow LS: Molecular basis for the auxin independent phenotype of crown gall tumor tissues. Science 231: 616-618, 1986.
97. Uchimiya H et al.: Expression of foreign gene in callus derived from DNA-treated protoplasts of rice (Oryza sativa L). Molec. Gen. Genet. 204: 204-207, 1986.
98. Vaeck M et al.: Transgenic plants protected from insect attack. Nature 328: 33-37, 1987.
99. Wicker M: Tumors. In: Cell and Tissue Culture in Forestry, vol. 2, JM Bonga and DJ Durzan (eds), Martinus Nijhoff Publishers, Dordrecht, pp. 374-389, 1987.
100. Yamada Y, Y Zhi-Qi and T Ding-Tai: Plant regeneration from protoplasts-derived callus of rice (Oryza sativa L.) Plant Cell Rep. 5: 85-88, 1986.
101. Zymbryski P et al.: Ti plasmid vector for the introduction of DNA into plant cells without alteration of their normal regeneration capacity. EMBO J. 2: 2143-2150, 1983.

SOMATIC CELL GENETIC RESEARCH IN FORESTRY: INTEGRATION OF CYTOGENETICS, TISSUE CULTURE, AND MOLECULAR GENETICS

S. E. SCHLARBAUM

DEPARTMENT OF FORESTRY, WILDLIFE AND FISHERIES
THE UNIVERSITY OF TENNESSEE
KNOXVILLE, TENNESSEE
UNITED STATES OF AMERICA 37901-1071

ABSTRACT

Somatic cell genetic research utilizes a combination of cytogenetic, tissue culturing, and molecular techniques to make detailed characterizations of human and mammalian genomes. In forestry, somatic cell genetic protocols can eliminate constraints that have inhibited basic research on genetic constitutions of trees. Mammalian research areas with pertinence to forest tree genetics are somatic cell hybridization, chromosome- and microcell-mediated gene transfer, and flow cytogenetics. Applications of mammalian procedures in plant genetic research are summarized, and the future use of these techniques for gene mapping in trees is elucidated. Potential applications of somatic cell genetic research in tree improvement include screening for desirable genotypes by genetic fingerprinting and parasexual synthesis of multi-parent trees.

1. INTRODUCTION

The development over the last three decades of somatic cell genetic research using human cell cultures has allowed researchers to explore the human genome to an extent previously not thought possible (54) (58) (81) (83). Utilization of cytogenetic, tissue culturing, and molecular techniques have produced detailed linkage maps of biochemical and genetic markers assigned to specific human chromosomes (2) (13) (19) (21) (35) (58) (60) (81) (90) (102). These linkage maps have been used as the basis for further studies in mutagenesis, gene regulation, genomic organization, and other facets of human and mammalian genetic research (82). The adoption of mammalian protocols into plant genetic research has the potential to significantly expand knowledge of genetic architecture, thereby increasing the possibilities and precision of genetic manipulations.

Initiation of somatic cell genetic research on forest tree species is of paramount importance as various constraints, e.g., long term reproductive cycles, inhibit research contributions in basic genetics. The potential for somatic cell research in forestry to address basic questions has been largely overlooked (1) (32) (59) (61) (88). This oversight was partially due to the general euphoria generated by the prospects of genetically engineering plants through parasexual procedures by protoplast fusion or plasmid-mediated transformation. Additionally the lack of emphasis on cytogenetic research of forest tree species also was a contributing factor (88).

A number of research areas in mammalian genetics are beginning to be paralleled by plant scientists, as advances occur in tissue culture,

M. R. Ahuja (ed.), Somatic Cell Genetics of Woody Plants, 103–118.

molecular genetics, and chromosome banding of plant species. Mammalian research areas include somatic cell hybridization, chromosome- and microcell-mediated gene transfer, and flow cytogenetics. Each area of research has been utilized to map genes onto specific chromosomes, in addition to providing the basis for other in vitro genetic research areas. In the subsequent text, procedures used in each area will be briefly reviewed, initially with mammalian species, and followed by examples of current plant research. This review will be followed by a discussion of the potential accomplishments somatic cell genetic research techniques could yield when applied to forest tree species.

2. REVIEW

2.1

Somatic cell hybridization.

In vitro fusion was noticed as early as 1909 (84). However, it was not until 1961 that cell fusion was used as a research tool to study human genetics (4). The discovery that inactivated Sendai virus promoted cell fusion accelerated the use of cell hybrids in the initial stages of mammalian somatic cell genetic research (47). This discovery was augmented by the report by Weiss and Green (101) of preferential loss of human chromosomes in human X mouse hybrids. They found that all human chromosomes were gradually lost in in vitro cell divisions with the exception of 1-3 chromosomes. Subsequent manipulations by various researchers have yielded hybrid cell lines containing one or more alien chromosomes or portions of chromosomes (58) (81) (83) (102).

Figure 1 illustrates the production of somatic cell hybrids. Fusion is promoted between 2 cell lines of different species; each cell line containing a different selectable gene

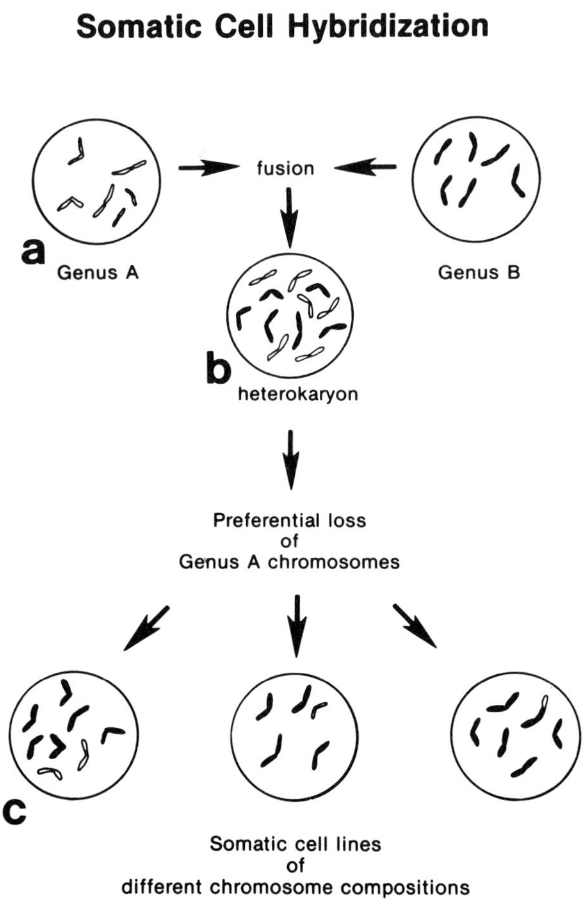

FIGURE 1. Production of somatic hybrid cell lines by cell hybridization.

(Fig. 1a). After an incubation period the cells are plated on a medium that allows only growth of heterokaryons containing both of the selectable genes (Fig. 1b). Cell lines are developed from the surviving heterokaryons and are examined for preferential loss of chromosomes and stability (Fig. 1c). Chromosome fragmentation, e.g., translocations and deletions, can occur after the cells have been kept in culture for a period of time (Fig. 1c-far right).

Contrasting biochemical analyses between parental and hybrid cell lines identify biochemical markers from the donor genome. In cell lines with alien chromosomes or chromosome fragments that can be identified by banding procedures, biochemical markers can be assigned to specific chromosomes or chromosome regions (58) (81) (83).

In plant species fusion of protoplasts have yielded heterokaryons that retain whole nuclear and/or chloroplast genomes of one donor species and partial genomes of the other donor species (3) (31) (36) (69) (74) (75) (77) (84) (91). An excellent example of alien chromosome transfer through protoplast fusion was published by Gupta et al. (43). The experiment was designed to transfer genes from Physalis minima to Datura innonxia. Albino Datura protoplasts were fused with normal Physalis protoplasts that had been irradiated. Green cell colonies were selected and plants regenerated from four selected transformed lines. Plants of two transformed lines were cytogenetically and biochemically characterized and contrasted with the donor species. They were found to have 48 Datura + 3 Physalis chromosomes and 96 Datura + 1 Physalis chromosomes, respectively. The presence of a portion of the Physalis genome was also confirmed by isoenzyme analysis that showed bands in the heterokaryons were present in both parents.

2.2 Chromosome- and microcell-mediated gene transfer.

Chromosome- and microcell-mediated gene transfers use similar techniques that transfer a portion of the nuclear genome from a donor to a recipient cell line. A successful experiment in gene transfer using isolated chromosomes was first published in 1973 by McBride and Ozer (70), who fused Chinese hamster chromosomes with mouse cells. Isolation and transfer of microcells, which are isolated micronuclei surrounded by a thin layer of cytoplasm, were independently reported by Ege and Ringertz (29) and Schor et al. (89) in 1975.

Figure 2 illustrates the methodology used in chromosome-mediated gene transfer (CMGT) and microcell-mediated gene transfer (MMGT). In each procedure a selectable gene is present in the donor genome. Mitosis is arrested in the donor cell line (Fig. 2a) to produce metaphase chromosomes in CMGT procedures or micronuclei in protocols of MMGT. The cells are then lysed, and the isolated chromosomes or microcells are purified from the debris (Fig. 2b). The donor chromosomes or microcells are mixed with the recipient cell line and fusion chemically promoted (Fig. 2c.). After an incubation period the cells are plated, and fusion products containing the selectable gene are identified by subsequent growth (Fig. 2d). These colonies are cultured as cell suspensions and are observed for stability of the alien donor material.

The end products of CMGT and MMGT are represented differently in Figure 2d. Cytologically identifiable alien chromosomes or chromosome fragments are present only in approximately 15 percent of the cell suspension lines resulting from CMGT procedures (72). The remaining 85 percent of the cell suspensions have donor chromatin integrated at the submicroscopic level. In contrast, MMGT experiments yield higher percentages of cell suspensions that contain whole alien chromosomes (2) (71) (87). Mapping of biochemical markers occurs by contrasting analysis of hybrid and parental cell lines coupled with chromosome analyses (if possible) as with somatic hybrids.

A limitation of CMGT and MMGT has been the selection of colonies that contain only genetic material associated with the selectable marker, either by linkage or co-organelle transfer. Fusion products involving chromosomes or microcells not containing the selectable marker are lost in the proliferation stage (Fig. 2d). This problem is being circumvented by molecular geneticists who are integrating selectable markers into different chromosomes (87).

Isolation of chromosomes has been reported in a number of horticultural and agronomic crop species (17) (22-25) (38-41) (44) (45) (67) (68) (79) (97). Incorporation of isolated chromosomes into recipient protoplasts has been reported by various researchers (38) (41) (79) (97). However, these studies were limited to observations of chromosome uptake and did not report on gene transfer. Recently, a complete study on chromosome-mediated gene transfer between plant species has been published. Griesbach (39) transferred genes from Petunia alpicola into Petunia hybrida by injecting isolated chromosomes into protoplasts. Confirmation of gene transfer was made by biochemical analyses of the resulting calli and transformed plants.

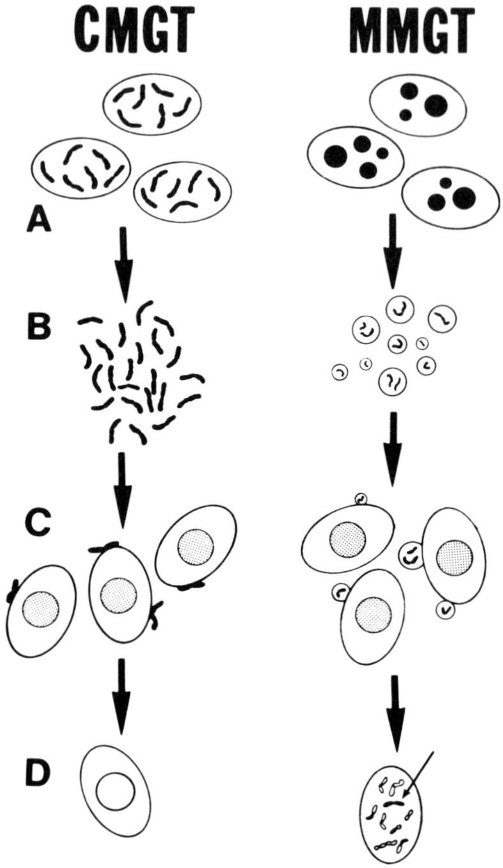

FIGURE 2. Chromosome transfer by chromosome-mediated gene transfer (CMGT) and microcell-mediated gene transfer (MMGT) techniques.

To date, microcell-mediated gene transfer has not been accomplished in plant species. Micronuclei have been induced in several plant species such as Allium (50-52). However, these studies were generally on cell cycle events rather than directed towards MMGT. Using the protocols developed for Allium, Schlarbaum (88) was able to induce micronuclei in Pinus strobus, but in a proportion of cells too low for MMGT research. Recently, Sree

Ramulu _et al_. (94) found that treating cell suspensions of _Nicotiana_, _Solanum_, _Haplopappus_, and _Daucus_ with amiprophos-methyl, an organophosphorous herbicide, would induce micronuclei in a relatively high proportion of cells. In the future, procedures for isolation of microcells will probably follow established techniques for isolation of plant nuclei (18) (76) (85) (86).

Various techniques are available for transferring isolated chromosomes and microcells to recipient cell lines. Chemicals (84) (91) and electricity (6) (106) (107), as well as microinjection (38) (39) (65), have been successfully utilized to fuse cells and/or transfer organelles in plant species.

2.3 Flow cytogenetics.

This area of research is a blend of classical cytogenetics and the modern mechanized technology of a cell sorter (9) (64) (105). Chromosomes are isolated from cell suspensions and stained with a DNA-specific fluorochrome. The suspension is placed in the cell sorter and then passed through the focusing point of a laser(s). The laser(s) excites the fluorescent dye causing light to be emitted proportional to the DNA content of each chromosome. The cell sorter operator can target certain chromosomes for sorting by their respective fluorescence. Sorted chromosomes are banded for specific identification and to ascertain the sorted suspension's purity. Chromosome identification by banding also allows the operator to correlate light intensities with specific chromosomes, i.e., a flow karyotype, so that banding does not have to be conducted in subsequent sorting experiments.

Sorting of isolated chromosomes was first reported by Gray _et al_. (37), who separated individual chromosomes derived from Chinese hamster cell suspensions. In the subsequent years, chromosomes of human and other mammalian species have been sorted (15) (16). The sorted chromosomes are used for a variety of genetic experiments including gene mapping (55) (63), construction of chromosome-specific DNA libraries (20) (46) (62), and studies on chromosome proteins (99) (104).

Previously, assignment of a genetic probe to a specific chromosome was accomplished by sorting of chromosomes into two fractions, digestion of DNAs with restriction enzymes, electrophoresis, blotting, and hybridization according to Southern (93). Fractionation and Southern hybridization were repeated until the single hybridizing chromosome or chromosome group was identified. However, this technique required a large amount of DNA. Currently a technique known as flow [spot] blotting is used to assign probes to a specific chromosome (9) (42) (66). In flow blotting the chromosomes are sorted directly onto a series of filters, based on the peaks of a flow karyotype. _In situ_ hybridizations are then conducted directly on the filter series. A detectable hybridization signal of a radioactive probe on a particular filter indicates the presence of that probe on the corresponding chromosome.

DNA libraries specific for different chromosomes can be constructed from sorted suspensions. Construction of the library involves sorting, isolation and purification of the DNA, restriction endonuclease digestion and insertion of DNA fragments into a plasmid or phage vector. Chromosome-specific libraries can be used for a variety of purposes in basic genetic research, e.g., sequence analysis, probes for restriction fragment length polymorphisms, etc. (62) (63) (95).

Sorting of chromosomes has been reported in only one plant species, *Haplopappus gracilis* (22) (23) (25) (79). The chromosomes were isolated from protoplasts derived from cell suspensions. Karyological analysis of the sorted suspensions revealed that separation of different chromosomes had occurred (23). Studies on the sorted chromosomes by electron microscopy and DNA analysis indicated that DNA integrity and general structure was preserved throughout the isolation and sorting procedures (25).

3. DISCUSSION

The above review of somatic cell genetic research in mammalian species exemplifies the utility of the techniques in providing basic genetic information that could not be obtained otherwise. The examples of research on plant species show the feasibility of applying mammalian protocols to plants. Current tissue culture and molecular genetic research on forest tree species is at a level of sophistication necessary for somatic cell genetic research (10-12). Cytogenetic procedures used on forest trees are behind state-of-the-art techniques, i.e., chromosome banding, developed in many agronomic and horticultural species (88). However, this is primarily a function of the lack of cytogenetic research on trees rather than any inherent problems. Therefore, somatic cell genetic research on tree species can be presently initiated with a reasonable probability of success.

3.1 Gene Mapping.

3.1.1 *Biochemical markers and genetic probes.* Applied tree improvement activities must be complemented by basic information to maintain consistency in future genetic improvement efforts. Mapping of biochemical markers and genetic probes should become an integral part in basic research on forest trees. Linkage maps have been used to initiate a myriad of basic genetic studies as alluded to in the Introduction. Futhermore, construction of genetic maps has relevance to splicing of genes by molecular techniques. Recognition of where an alien gene is spliced into a recipient tree's genome is important as integration into different linkage groups produces different results in the transformed plants (33) (100).

Biochemical marker linkage groups have been constructed in tree species but not assigned to specific chromosomes (27) (30). Initiation of somatic cell genetic research using hybrid cell suspensions produced by fusion, CMGT, and MMGT procedures will make possible assignment of these linkage groups to specific chromosomes and chromosome regions. Manipulations of flow sorted chromosomes will provide genetic probes for chromosome assignment and use in other molecular research. Procedures are available now for direct *in situ* hybridization of genetic probes onto metaphase chromosomes in plants (5) (34) (56) (57) (80).

3.1.2 *Restriction fragment length polymorphisms (RFLP).* Research on this class of genetic markers has been progressively increasing since their utility was demonstrated in human research (13) (26) (35) (53). Restriction fragment length polymorphisms provide opportunities to develop detailed genetic maps, starting points for gene cloning, and possible correlations with quantitative traits (7) (14) (48) (49) (73) (78) (92). Restriction fragment length polymorphism analysis frequently utilizes radioactive labeled probes to identify genomic variation at the DNA level. This variation is expressed as length differences in DNA fragments released by restriction enzyme digestions. Fragment polymorphism can result from base changes that affect the length between cleavage sites. The polymorphisms are visualized on ethidium bromide stained gels or an autoradiograph after the fragments have been blotted and hybridized.

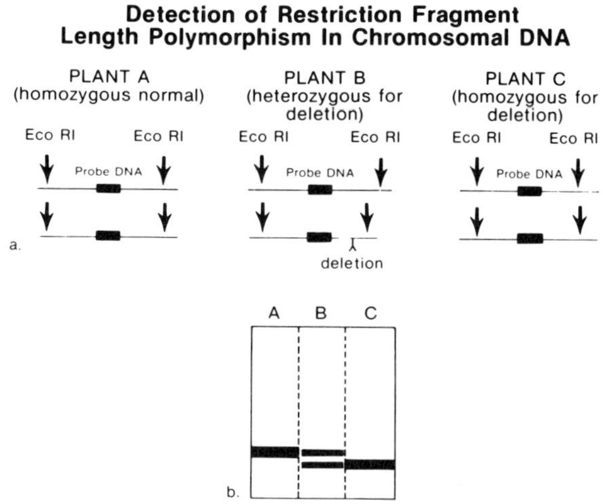

FIGURE 3. Restriction fragment length polymorphism analysis.

Nuclear RFLPs are inherited in a codominant Mendelian mode. An example of RFLP detection in chromosomal DNA is shown in Figure 3. The nuclear DNA of Plants A, B, and C (diploid state assumed) is digested with a restriction endonuclease, Eco RI, and the resulting fragments hybridized with a selected radioactive probe (Fig. 3a). In Plant A the fragments are identical in size (Fig. 3a-left), and the corresponding autoradiograph shows only one band (Fig. 3b-left). Plant B is heterozygous for a deletion in the fragment regions (Fig. 3a-middle), and the corresponding autoradiograph shows two bands (Fig. 3b-middle). Fragments in Plant C are homozygous for the deletion (Fig. 3a-right) and are exhibited in the autoradiograph as one band (Fig. 3b-right).

Sorted chromosomes and pedigree information can be used in RFLP analysis. Probes made from chromosome-specific DNA libraries have been used to detect RFLPs in humans (53) (95). Starr and Wood (95) hybridized radioactive probes of the X-chromosome short arm to Southern blots of genomic digests. Three probes were found to detect RFLPs. Pedigree analysis showed that the inheritance of restriction fragments was linked to the X-chromosome.

A possible approach to constructing chromosome-specific RFLP maps is shown in Figure 4. Chromosomes are sorted from plant material or cell suspensions (Fig. 4a). A chromosome-specific library is constructed in lambda phage vectors (Fig. 4b) and screened for unique or low copy sequences. Chromosome-specific probes are selected from the pool of unique and low copy sequences. The probes can be used to detect RFLPs in sorted chromosomes that have been digested with restriction enzymes, electrophoresed on gels, blotted, and hybridized. Figure 4c illustrates hybridization results of 2 different probes to specific chromosomes. Hybridizations involving chromosome 1 indicates that the sorted chromosomes were homozygous for a RFLP. The two bands in the hybridization involving chromosome **n** shows that the sorted chromosomes were heterozygous for a different RFLP. The use of probes from chromosome-specific DNA libraries derived from hybrid cell lines that contain aberations (103), e.g., deletions and translocations, will allow for assignment of RFLPs to chromosome arms or specific regions (Fig. 4d).

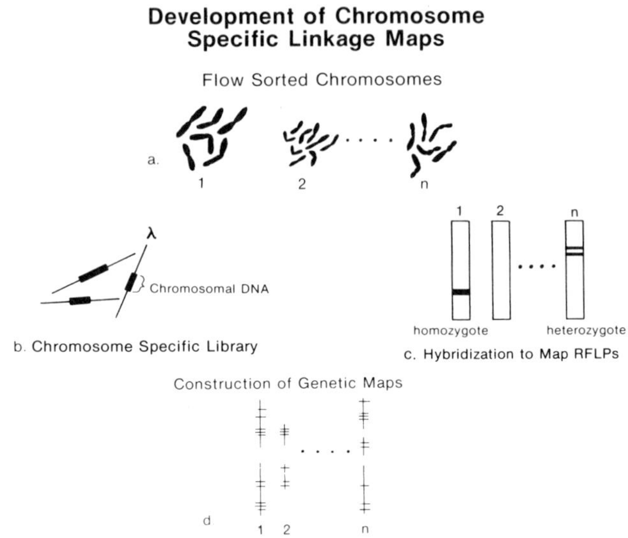

FIGURE 4. Theoretical representation of restriction fragment length polymorphism analysis using flow sorted chromosomes.

3.1.3 <u>Saturated linkage maps</u>. Information on the chromosomal location of biochemical members, genetic probes, and RFLPs from somatic cell genetic procedures will produce saturated linkage maps for each chromosome type (8). Figure 5 is a hypothetical representation of linkage group assignments of various markers to physical locations on banded chromosome 5 of *Cunninghamia konishii*.

3.2 <u>Somatic cell genetics and tree improvement</u>. The basic information on genetic constitutions of trees obtained from somatic cell genetics can be utilized in applied tree improvement activities. Field testing of experimental materials in forestry is both time and space consuming. Correlation of genetic markers with important quantitative and qualitative traits (7) (14) (28) (48) (92) (96) (98) will allow scientists to detect trees with desirable qualities, thereby reducing the size

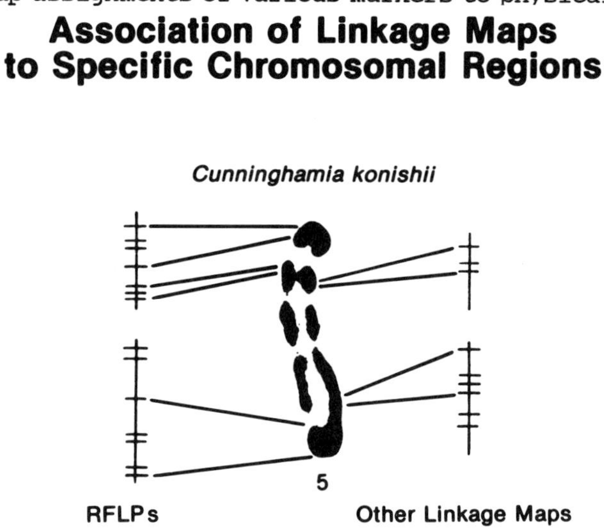

FIGURE 5. Representation of a saturated linkage map assigned to chromosome 5 of *Cunninghamia konishii*.

of experimental plantations. By combining this information with breeding, screening for desirable genes and gene linkages by biochemical and/or DNA analyses (genetic fingerprinting) could be possible as in Figure 6. In certain circumstances, high correlations between genetic fingerprints and overall field performance may allow for the elimination of field-based experiments and incorporation of screened genotypes directly into the developmental phases of tree improvement. Genetic fingerprinting will also detect recombination events in genetic materials. It would be highly desirable to have trees with important genes, quantitative and/or qualitative, that are tightly linked and inherited as a single unit in a Mendelian fashion.

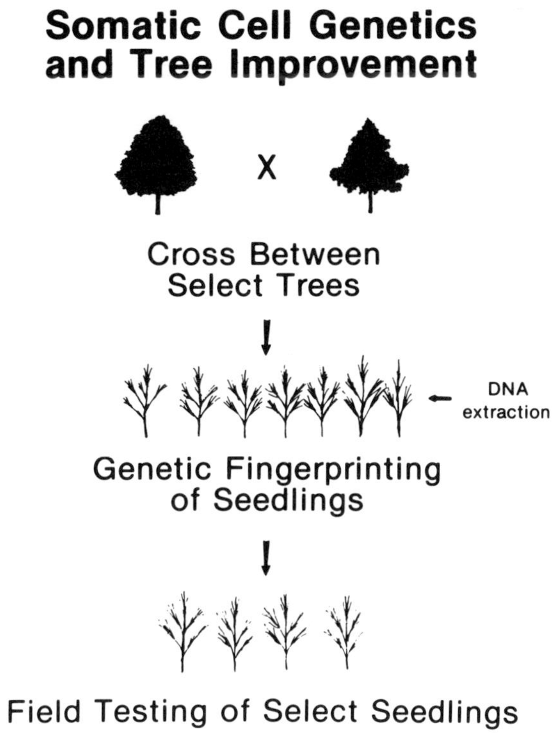

FIGURE 6. Incorporation of genetic fingerprinting into tree improvement research.

Genetic engineering of forest trees using parasexual procedures should be directed towards recombining quantitative in addition to qualitative traits. Many of the important qualities in commercial forest trees are quantitatively inherited. Research also should be conducted on intraspecific as well as interspecific recombination. The long term reproductive cycles of most forest trees prevent rapid recombination of desirable genes or gene linkages into different genetic backgrounds within a species. Furthermore, genes successfully integrated by molecular techniques need to be recombined into different genotypes. Accelerated breeding techniques can aid in rapid recombination among trees with desirable genotypes. However, these techniques generally require a number of years before sexual maturation of the breeding materials and up to 2 growing seasons for maturation of seed from controlled crosses.

Information on genomic constitutions from somatic cell genetic studies, coupled with flow sorting of chromosomes and tissue culture, may provide an opportunity for rapid recombination of quantitative and qualitative traits in vitro. As important genes are mapped onto individual chromosomes, certain chromosomes or chromosome groups can be identified as containing the gene(s) controlling or predominantly controlling a characteristic. It may be desirable to combine certain intact portions of the chromosome complements of various trees.

A hypothetical genomic reconstitution of trees at the chromosome level is shown in Figure 7. Certain chromosomes that contain desirable genes or linkage groups are flow sorted from 3 different trees within a species. As shown in Figure 7, the chromosome complement of the species, 2n=24 chromosomes with twelve homologous pairs, is reconstituted with different contributions from the 3 donor trees. The chromosome pairs are fused or microinjected into an enucleated protoplast. The protoplast is then regenerated into a 2n=24 seedling that has a blend of desirable characteristics from the 3 donor trees. This technique also has potential to create alien chromosome addition or substitution lines as developed in agronomic and horticultural crop species through breeding.

Integration of Cytogenetics, Molecular Genetics and Tissue Culture For Tree Improvement

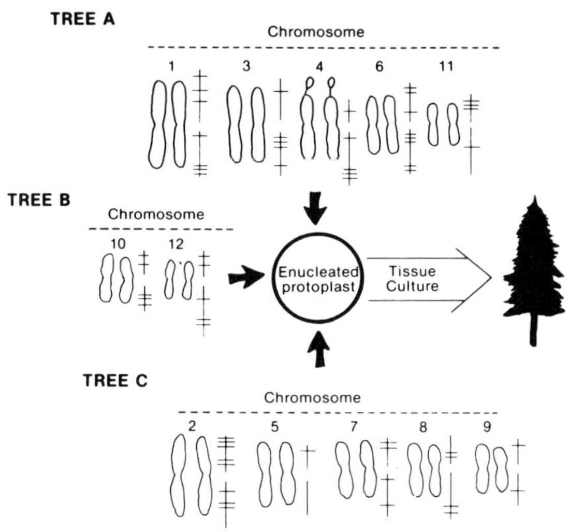

FIGURE 7. Theoretical representation of the synthesis of a multi-parent tree using sorted chromosomes from 3 donor trees.

4. CONCLUDING REMARKS

Progressive research in cytogenetics, molecular genetics, and tissue culture of forest trees increases the probability for successful somatic cell genetic studies. By paralleling human and mammalian techniques, researchers can begin to explore genomic organization and genetic constitutions of trees with increased precision. Study results will provide basic information on genetic mechanisms in trees and potentially have ramifications for tree improvement. Close cooperation between basic and applied research efforts in tree genetics should greatly expand the knowledge base in a relatively short period of time and have a significant input on commercial forestry in the near future.

REFERENCES

1. AHUJA MR: Protoplast research in woody plants. Silvae Genet 33: 32-37, 1984.
2. ATHWAL RS, V DHAR: Transfer of single specific chromosomes by microcell fusion and gene mapping in mammalian cells. In ML Pearson, NL Sternberg eds, Gene Transfer and Cancer. New York: Raven Press, pp 21-29, 1984.
3. AVIV D, P ARZEE-GONEN, S BLEICHMAN, E GALUM: Novel alloplasmic Nicotiana plants by "donor-recipient" protoplast fusion: hybrids having N. tabacum or N. sylvestris nuclear genomes and either or both plastomes and chondriomes from alien species. Mol Gen Genet 196: 244-453, 1984.
4. BARSKI G, S SORIEUL, F CORNEFERT: "Hybrid" type cells in combined cultures of two different mammalian cell strains. J Nat Cancer Inst 26: 1269-1291, 1961.
5. BARTELS D, I ALTOSAAR, NP HARBERD, RF BARKER, RD THOMPSON: Molecular analysis of y-gliadin gene families at the complex Gli-1 locus of bread wheat (T. aestivum L). Theor Appl Genet 72: 845-853, 1986.
6. BATES GW, CA HASENKAMPF: Culture of plant somatic hybrids following electric fusion. Theor Appl Genet 70: 227-233, 1985.
7. BECKMAN, J.S. AND M. SOLLER: Restriction fragment length polymorphisms and genetic improvement of agricultural species. Euphytica 35: 111-124, 1986.
8. BERNATZKY R, SD TANKSLEY: Toward a saturated linkage map in tomato based on isozymes and random cDNA sequences. Genetics 112: 887-898, 1986.
9. BERNHEIM, A. AND R. MIGLIERINA: Flow cytogenetics. Biol. Cell 58: 179-182, 1986.
10. BONGA JM, DJ DURZAN (eds): Cell and Tissue Culture in Forestry. Volume 1. General Principles and Biotechnology. Dordrecht: Martinus Nijhoff Publishers, 1987.
11. BONGA JM, DJ DURZAN (eds): Cell and Tissue Culture in Forestry. Volume 2. Specific Principles and Methods: Growth and Development. Dordrecht: Martinus Nijhoff Publishers, 1987.
12. BONGA JM, DJ DURZAN (eds): Cell and Tissue Culture in Forestry. Volume 3. Case Histories: Gymnosperms, Angiosperms and Palms. Dordrecht: Martinus Nijhoff Publishers, 1987.
13. BOTSTEIN D., R.L. WHITE, M. SKOLNICK AND R.W. DAVIS: Construction of a genetic linkage map in man using restriction fragment length polymorphisms. Am J Hum Genet 32: 314-331. 1980.
14. BURR, B., S.V. EVOLA, F.A. BURR AND J.S. BECKMANN: The application of restriction fragment length polymorphism to plant breeding. In JK Setlow, A Hollaender eds Genetic Engineering Principles and Methods. New York Plenum Press Vol. 5: 45-59, 1983.
15. CARRANO AV, JW GRAY, RG LANGLOIS, KJ BURKHART-SCHULTZ, MA VAN DILLA: Measurement and purification of human chromosomes by flow cytometry and sorting. Proc Nat Acad Sci USA 76: 1382-1384, 1979.
16. CARRANO AV, JW GRAY, DH MOORE II, JL MINKLER, BH MAYALL, MA VAN DILLA, ML MENDELSOHN: Purification of the chromosomes of the Indian muntjac by flow sorting. J Histochem Cytochem 24: 348-354, 1976.

17. CHAPMAN GP, SA COOKE: A technique for optical and electron microscopy of isolated plant chromosomes. Protoplasma 116: 198-200, 1983.
18. CHAPPELL J, K HAHLBROCK: Salt effects on total and gene-specific in vitro transcriptional activity of isolated plant nuclei. Plant Cell Rep 5: 398-402, 1986.
19. CHU EHY, SS POWELL: Selective systems in somatic cell genetics. Adv Hum Genet 7: 189-258, 1976.
20. DAVIES KE, BD YOUNG, RG ELLES, ME HILL, R WILLIAMSON: Cloning of a representative genomic libary of the human X chromosome after sorting by flow cytometry. Nature 293: 374-376, 1981.
21. DE JONGE AJR, D BOOTSMA: Chromosome and DNA-mediated gene transfer in cultured mammalian cells. Int Rev Cytol 92: 133-158, 1984.
22. DE LAAT AMM, J BLAAS: Isolation, purification and sorting by flow cytometry of metaphase chromosomes of _Haplopappus gracilis_. Poster, Proc Sixth Int Protoplast Symp. pp 121-122, 1983.
23. DE LAAT AMM, J BLAAS: Flow-cytometric characterization and sorting of plant chromosomes. Theor Appl Genet 67: 463-467, 1984.
24. DE LAAT AMM, J BLAAS: An improved method for protoplast micro-injection suitable for transfer of entire plant chromosomes. Plant Sci 50: 161-169, 1987.
25. DE LAAT AMM, JHN SCHEL: The integrity of metaphase chromosomes of _Haplopappus gracilis_ (Nutt.) Gray isolated by flow cytometry. Plant Sci 47: 145-151, 1986.
26. DRAYNA, D., K. DAVIS, D. HARTLEY, J.-L. MANDEL, G. CAMIRINO, R. WILLIAMSON AND R. WHITE: Genetic mapping of the human X-chromosome by using restriction fragment length polymorphisms. Proc Natl Acad Sci USA 81: 2836-2839, 1984.
27. ECKERT RT, RJ JOLY, DB NEALE: Genetics of isozyme variants and linkage relationships among allozyme loci in 35 eastern white pine clones. Can J For Res 11: 573-579, 1981.
28. EDWARDS, M.D., C.W. STUBER AND J.F. WENDEL: Molecular-marker-facilitated investigations of quantitative-trait loci in maize. I. Numbers, genomic distribution and types of gene action. Genetics 116: 113-125, 1987.
29. EGE T, NR RINGERTZ: Preparation of microcells by enucleation of micronucleate cells. Exp Cell Res 87: 378-382, 1974.
30. EL-KASSABY YA, O SZIKLAI, FC YEH: Linkage relationships among 19 polymorphic allozyme loci in coastal Douglas-fir (_Pseudotsuga menziesii_ var. _menziesii_). Can J Genet Cytol 24: 101-108, 1982.
31. EVANS DA: Agricultural applications of plant protoplast fusion. Bio/Tech 1: 253-261, 1983.
32. FARNUM P, R TIMMIS, JL KULP: Biotechnology of forest yield. Science 219: 694-702, 1983.
33. FRALEY RT, SG ROGERS, RB HORSCH: Genetic transformation in higher plants. Crit Rev Plant Sci 4: 1-46, 1986.
34. GERLACH WL, WJ PEACOCK: Chromosomal locations of highly repeated DNA sequences in wheat. Heredity 44: 269-276, 1980.
35. GOATE, A.M., D.N. COOPER, C. HALL, T.K.C. LEUNG, E. SOLOMAN AND L. LIM: Localization of a human heat-shock HSP 70 gene sequence to chromosome 6 and detection of two other loci by somatic-cell hybrid and restriction fragment length polymorphism analysis. Hum Genet 75:

123-128, 1987.

36. GOODMAN RM, H HAUPTLI, A CROSSWAY, VC KNAUF: Gene transfer in crop improvement. Science 236:48-54, 1986.
37. GRAY JW, AV CARRANO, LL STEINMETZ, MA VAN DILLA, DH MOORE II, BH MAYALL, ML MENDELSOHN: Chromosome measurement and sorting by flow systems. Proc Nat Acad Sci USA 72: 1231-1234, 1975.
38. GRIESBACH RJ: Protoplast microinjection. Plant Mol Biol Rep 1: 32-37, 1983.
39. GRIESBACH RJ: Chromosome-mediated transformation via microinjection. Plant Sci 50: 69-77, 1987.
40. GRIESBACH RJ, RL MALMBERG, PS CARLSON: An improved technique for the isolation of higher plant chromosomes. Plant Sci Lett 24: 55-60, 1982.
41. GRIESBACH RJ, RL MALMBERG, PS CARLSON: Uptake of isolated lily chromosomes by tobacco protoplasts. J Hered 73: 151-152, 1982.
42. GRUNWALD D, C GEFFROTIN, P CHARDON, G FRELAT, M VAIMAN: Swine chromosomes: flow sorting and spot blot hybridization. Cytometry 7: 582-588, 1986.
43. GUPTA PP, O SCHIEDER, M GUPTA: Intergeneric nuclear gene transfer between somatically and sexually incompatible plants through asymmetric protoplast fusion. Mol Gen Genet 197: 30-35, 1984.
44. HADLACZKY G, T PRAZNOVSZKY, G BISZTRAY: Structure of isolated protein-depleted chromosomes of plants. Chromosoma 86: 643-659, 1982.
45. HADLACZKY G, G BISZTRAY, T PRAZNOVSZKY, D DUDITS: Mass isolation of plant chromosomes and nuclei. Planta 157: 278-285, 1983.
46. HARRIS P, E BOYD, MA FERGUSON-SMITH: Optimising human chromosome separation for the production of chromosome-specific DNA libraries by flow sorting. Hum Genet 70: 59-65, 1985.
47. HARRIS H, JF WATKINS: Hybrid cells derived from mouse and man: artificial heterokaryons of mammalian cells from different species. Nature 205: 640-646, 1965.
48. HELENTJARIS T, G. KING, M. SLOCUM, C. SIEDEMSTRANG AND S. WEGMAN: Restriction fragment polymorphisms as probes for plant breeding. Plant Mol Biol 5: 109-118, 1985.
49. HELENTJARIS T, M. SLOCUM, S. WRIGHT, A. SCHAEFER AND J. NIENHUIS: Construction of genetic linkage maps in maize and tomato using restriction fragment length polymorphisms. Theor Appl Genet 72: 761-769, 1986.
50. HERVAS JP: Multinucleate plant cells I. Aneuploidy in a proliferating population. Exp Cell Res 97: 203-212, 1976.
51. HERVAS JP, JF LOPEZ-SAEZ, G GIMENEZ-MARTIN: Multinucleate plant cells II. Requirements for nuclear DNA synthesis. Exp Cell Res 139: 341-350, 1982.
52. HERVAS JP, JF LOPEZ-SAEZ, G GIMENEZ-MARTIN: Proliferation of multinucleate cells with unbalanced nuclei. Cell Biol Int Rep 9: 91-101, 1985.
53. HOFKER, M.H., M.C. WAPENAAR, N. GOOR, E. BAKKER, G.-J.B. VAN OMMEN AND P.L. PEARSON: Isolation of probes detecting restriction fragment length polymorphisms from X chromosome specific libraries: potential use for diagnosis of Duchenne muscular dystrophy. Hum. Genet., 70: 148-156, 1985.
54. HSU TC: Human and Mammalian Cytogentics. A Historical Perspective.

New York: Springer-Verlag New York Inc, 1979.
55. HUGHES SH, E STUBBLEFIELD, F PAYVAR, JD ENGEL, JB DODGSON, D SPECTOR, B CORDELL, RT SCHIMKE, HE VARMUS: Gene localization by chromosome fractionation: globin genes are on at least two chromosomes and three estrogen-inducible genes are on three chromosomes. Proc Nat Acad Sci USA 76: 1348-1352, 1979.
56. HUTCHINSON J and AG SEAL: A sequential in-situ hybridization and C-banding technique. Heredity 51: 507-509, 1983.
57. HUTCHINSON J, RB FLAVELL and J JONES: Physical mapping of plant chromosomes by in-situ hybridization. In JK Setlow, A Hollaender eds, Genetic Engineering Principles and Methods. New York: Plenum Press, 3: 207-222, 1982.
58. KAO FT: Somatic cell genetics and gene mapping. Int Rev Cytol 85: 109-146, 1983.
59. KARNOSKY DF: Potential for tree improvement via tissue culture. BioSci 31: 114-120, 1981.
60. KLOBUTCHER LA, FH RUDDLE: Chromosome mediated gene transfer. Annu Rev Biochem 50: 533-554, 1981.
61. KRIEBEL HB: Research approaches to forest genetics at the DNA level. Proc 6th North Am For Biol Workshop, pp 49-76, 1980.
62. KRUMLAUF R, M JEANPIERRE, BD YOUNG: Construction and characterization of genomic libraries from specific human chromosomes. Proc Nat Acad Sci USA 79: 2971-2975, 1982.
63. KUNKEL LM, U TANTRAVAHI, M EISENHARD, SA LATT: Regional localization on the human X of DNA segments cloned from flow sorted chromosomes. Nucleic Acid Res 10: 1557-1578, 1982.
64. LANGLOIS, R.G., L.-C. YU, J.W. GRAY AND A.V. CARRANO: Quantitative karyotyping of human chromosomes by dual beam flow cytometry. Proc Natl Acad Sci USA 79: 7876-7880, 1982.
65. LAWRENCE WA, DR DAVIES: A method for the microinjection and culture of protoplasts at very low densities. Plant Cell Rep 4: 33-35, 1985.
66. LEBO RV, F GORIN, RJ FLETTERICK, FT KAO, MC CHEUNG, BD BRUCE, AND YW KAN: High resolution chromosome sorting and DNA spot-blot analysis assign McArdle's syndrome to chromosome 11. Science 225: 57-59, 1984.
67. MALMBERG RL, RJ GRIESBACH: The isolation of mitotic and meiotic chromosomes from plant protoplasts. Plant Sci Lett 17: 141-147, 1980.
68. MATTHEWS BF: Isolation of mitotic chromosomes from partially synchronized carrot (D. carota) cell suspension cultures. Plant Sci Lett 31: 165-172, 1983.
69. MATTHEWS BF, JM WIDHOLM: Organelle DNA compositions and isoenzyme expression in an interspecific somatic hybrid of Daucus. Mol Gen Genet 198: 371-379, 1985.
70. MCBRIDE OW, HL OZER: Transfer of genetic information by purified metaphase chromosomes. Proc Nat Acad Sci USA 70: 1258-1262, 1973.
71. MCNEIL CA, RL BROWN: Genetic manipulation by means of microcell-mediated transfer of normal human chromosomes into recipient mouse cells. Proc Nat Acad Sci USA 77: 5394-5398, 1980.
72. MURPHY DD, CL MILLER, FH RUDDLE: Quantitation of the transgenome size in chromosome-mediated gene transfer lines. Cytogenet Cell Genet 39: 125-133, 1985.
73. NIENHUIS J,T HELENTJARIS, M SLOCUM, B RUGGERO, A SCHAEFER: Restriction fragment length polymorphism analysis of loci associated

with insect resistance in tomato. Crop Sci 27: 797-803, 1987.
74. PENTAL D, JD HAMILL, A PIRRIE, EC COCKING: Somatic hybridization of Nicotiana tabacum and Petunia hybrida. Mol Gen Genet 202: 342-347, 1986.
75. PIJNACKER LP, MA FERWERDA, KJ PUITE, S ROEST: Elimination of Solanum phureja nucleolar chromosomes in S. tuberosum + S. phureja somatic hybrids. Theor Appl Genet 73: 878-882, 1987.
76. POTRYKUS I, F HOFFMAN: Transplantation of nuclei into protoplasts of higher plants. Z Pflanzenphysiol 69: 287-289, 1973.
77. POTRYKUS I, J JIA, GB LAZAR, M SAUL: Hyoscyamus muticus + Nicotiana tabacum fusion hybrids selected via auxotroph complementation. Plant Cell Rep 3: 68-71, 1984.
78. POULANS, N.O., N.F. WEEDEN AND W.F. THOMPSON: Variation in pea restriction endonuclease fragment patterns. Genetics 107: s84, 1984.
79. PUITE KJ, LJW GILISSEN, AMM DE LAAT: First steps to chromosome transplantation with Haplopappus gracilis using flow cytometry. Poster, Proc Sixth Int Protoplast Symp. pp 123-124, 1983.
80. RAYBURN AL, BS GILL: Use of biotin-labeled probes to map specific DNA sequences on wheat chromosomes. J Hered 76: 78-81, 1985.
81. RUDDLE FH: Linkage analysis using somatic cell hybrids. Adv Hum Genet 3: 173-235, 1972.
82. RUDDLE FH: A new era in mammalian gene mapping: somatic cell genetics and recombinant DNA methodologies. Nature 294: 115-120, 1981.
83. RUDDLE FH, RP CREAGAN: Parasexual approaches to the genetics of man. Annu Rev Genet 9: 407-486, 1975.
84. SAUNDERS JA, GW BATES: Chemically induced fusion of plant protoplasts. In AE SOWERS ed, Cell Fusion. New York and London: Plenum Press, 1987.
85. SAXENA, PK, LC FOWKE, J KING: An efficient procedure for isolation of nuclei from plant protoplasts. Protoplasma 128: 184-189, 1985
86. SAXENA PK, Y LIU, J KING: Nuclear transplantation into plant protoplasts: optimal conditions for induction and determination of nuclear uptake. J Plant Physiol 128: 451-460, 1987.
87. SAXON, PJ, ES SRIVATSAN, GV LEIPZIG, JH SAMESHIMA, EJ STANBRIDGE: Selective transfer of individual human chromosomes to recipient cells. Mol Cell Biol 5: 140-146, 1985.
88. SCHLARBAUM SE: Cytogentic mamipulations in forest trees through tissue culture. In J Bonga, DJ Durzan eds, Tissue Culture in Forestry. Volume 1. General Principles and Biotechnology. Dordrecht: Martinus Nijhoff Publishers, pp. 330-352, 1987.
89. SCHOR SL, RT JOHNSON, AM MULLINGER: Perturbation of mammalian cell division II. Studies on the isolation and characterization of human mini segregant cells. J Cell Sci 19: 281-303, 1975.
90. SHOWS TB, AY SAKAGUCHI: Gene transfer and gene mapping in mammalian cells in culture. In Vitro 16: 55-76, 1980.
91. SINK KC: Protoplast fusion for plant improvement. HortSci 19: 33-37, 1984.
92. SOLLER, M. AND J.S. BECKMANN: Restriction fragment length polymorphisms and animal genetic improvement. In JSF Baker ed, Reviews in Rural Sciences. 6. Univ. New England, Armidale, Australia. pp 10-18. 1985.
93. SOUTHERN EM: Detection of specific sequences among DNA fragments

separated by gel electrophoresis. J Mol Biol 98: 503-517, 1975.

94. SREE RAMULA K, HA VERHOEVEN, P DIJKHUIS, LJWJ GILISSEN: Telophasing of metaphase chromosomes and formation of micronuclei after treatment of plant cells with amiprophos-methyl. Poster presentation, Kew Chromosome Conference III, 1987.

95. STARR, T. AND S. WOOD: Isolation and characterization of DNA probes from the short arm of the human X chromosome that detect restriction fragment length polymorphisms. Genome 29: 201-205, 1987.

96. STUBER, C.W., M.D. EDWARDS AND J.F. WENDEL: Molecular-marker-facilitated investigations of quantitative-trait loci in maize. II. Factors influencing yield and its component traits. Crop Sci 27: 639-648, 1987.

97. SZABADOS L, G HADLACZKY, D DUDITS: Uptake of isolated plant chromosomes by plant protoplasts. Planta 151: 141-145, 1981.

98. TANKSLEY, S.: Molecular markers in plant breeding. Plant Mol Biol Rep 1: 3-8, 1983.

99. TRASK B, G VAN DEN ENGH, J GRAY, M VANDERLAAN, B TURNER: Immunofluorescent detection of histone 2B on metaphase chromosome using flow cytometry. Chromosoma 90: 295-302, 1984.

100. WALLROTH M, AGM GERATS, SG ROGERS, RT FRALEY, RB HORSCH: Chromosomal localization of foreign genes in Petunia hybrida. Mol Gen Genet 202: 6-15, 1986.

101. WEISS MC, H GREEN: Human-mouse hybrid cell lines containing partial complements of human chromsomes and functioning human genes. Proc Nat Acad Sci USA 58: 1104-1111, 1967.

102. WIEACKER, P., K.E. DAVIS, H.J. COOKE, P.L. PEARSON, R. WILLIAMSON, S. BHATTACHARYA, J. ZIMMER AND H.-H. ROPERS: Toward a complete linkage map of the human X chromosome: regional assignment of 16 cloned single-copy DNA sequences employing a panal of somatic cell hybrids. Am J Hum Genet 36: 265-276, 1984.

103. WIRSCHUBSKY, Z., C. PERLMANN, J. LINDSTERN AND G. KLEIN: Flow karyotype analysis and fluorescence activated sorting of burkitt-lymphoma-associated translocation chromosomes. Int J Cancer 32: 147-153, 1983.

104. WRAY W, VP WRAY: Isolated metaphase chromosomes II. Proteins of Chinese hamster chromosomes. J Histochem Cytochem 27: 454-457, 1979.

105. YU LC, JW GRAY, R LANGLOIS, MA VAN DILLA, AV CARRANO: Human chromosome karyotyping and molecular biology by flow cytometry. In RS Sparks, FF de la Cruz, eds, Research Perspectives in Cytogenetics. Baltimore: University Park Press, pp 63-73, 1984.

106. ZACHRISSON Z, CH BORNMAN: Electromanipulation of plant protoplasts. Physiol Plant 67: 507-516, 1986.

107. ZIMMERMAN U, J VIENKEN: Electric field-induced cell-to-cell fusion. J Membr Biol 67: 165-182, 1982.

DIFFERENTIAL NORMS OF REACTION IN TISSUE CULTURE OF BIRCH

HEIDE GLOCK

Abteilung für Forstgenetik und Forstpflanzenzüchtung
der Universität
Büsgenweg 2, D-3400 Göttingen

Abstract

Diploid callus cultures of four birch genotypes are used for examining the interaction of the genotypic and the environmental effects on forming the phenotype. The characters observed were growth and colour of the calli after culture under eight different light conditions for six weeks on a modified MS-medium (Murashige and Skoog 1962).

The results show, that under the applied conditions the genotypic effects have a stronger influence on the phenotype than the environmental effects have. They depend on the genotypes, so that at least one-sided separability of the genotypic from the environmental effects could be observed.

Introduction

The technique of tissue culture is a very suitable object for obtaining some more precise information on genotype-environment-interactions. First of all there is the possibility to create defined environmental conditions and to use the same genotype in its repetitions, which have as nearly as possible the same ontogenetic state.

Therefore I will now try to demonstrate some aspects of the genotype environment interaction using tissue cultures of four birch genotypes cultivated in eight different light conditions. I will use the main ideas of a rather new concept develloped by Gregorius and Namkoong (1986). This method allows very specific and exact assertions about the single effects of the genotype and of the environment shaping the character under consideration.

Material

The experiment was conducted with diploid callus cultures of birch. The original tissue was cambium of internodes from 5 to 6 years old trees, denoted by G7, G8, G9, G11. These four individuals showed genealocical relationships.

M. R. Ahuja (ed.), Somatic Cell Genetics of Woody Plants, 119–125.

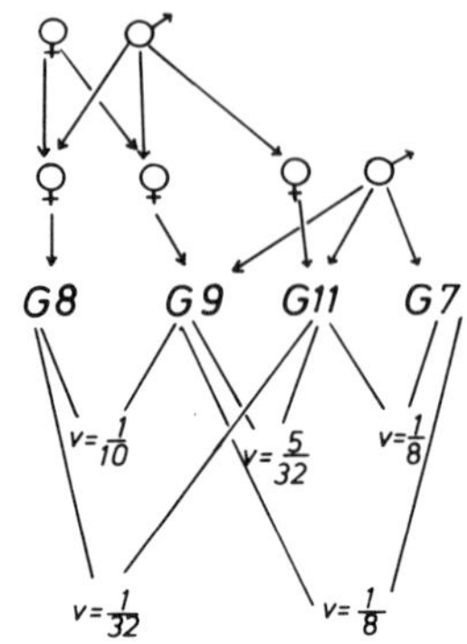

Fig. 1: Coefficients of kinship

G7, G9 and G11 are half sibs, with a common pollen parent, a red birch from the Arboretum in Göttingen. G9 and G11 are additionally related via their female parents: these are half sibs with a common pollen parent. G8 and G9 as well as G8 and G11 have no common parent but the female parents of G8 and G9 are full sibs and the female parents of G8 and G11 are half sibs via their pollen parent. G7 and G8 have no relationship. The coeffients of kinship are given on the foil.

Methods

Each of the genotypes was grown in the same range of environments. They were cultivated on a modified Murashige-Skoog medium; the pH was adjusted of 5.8 before autoclaving. The temperature was 25°-28° C and the air humidity was about 60% .

The environmental factor "light" was varied in its qualities blue red white and dark. After three weeks half of the samples in the blue were put in the red light (marked as BR) for another three weeks and vice versa (RB). The same was done with samples in the white and in the dark. So that there were eight light conditions tested.

Characters

The characters observed are "callus growth" and "callus colour".

The graphs of callus growth represent the volume increase of the calli related to that at the beginning of the experiment. Therefore each dot in the response function shows the caculated multiple of this initial volume.

Callus colour is determined as green, white, yellow and brown. The measurement of this character is the relative frequency of green calli among all of the same genotype and grown in the same environment. Brown calli are classified as not living. White and yellow calli are indeed not able to synthesize chlorophyll but are still living.

Model

For the interpretation of the results given later I adapted the model of Gregorius and Namkoong which allows to interpret any results in terms of the cause-effect-principle. In this case here, particular emphasis is put on the genotype environment interaction.

The basis of the concept is characterized by the term **"distinguishability of genotypic and environmental effects"**. The effects are discribed with the help of response functions. The response function of a particular genotype or its "norm of reaction" specifies the reaction of this genotype to different environmental conditions for a given trait variable. Similarly, the response function of an environmental factor specifies the effect this environmental situation has on different genoptypes.

Thus, if two genotypes differ completely in their responses to the environmental conditions tested, such that one genotype always ranks above the other over the whole set of environments, than their effects are said to be consistently distinguishable. Otherwise, if two genotypes have identical norms, their effects are identical. If either of these situations is true for a particular genotype relative to all other genotypes under consideration than the effect of this genotype is called to be consistent relative to this set of genotypes.If all genotypes have consistent effects than we speek of **separability of genotypic effects**. Analogously, the same is true considering environmental effects when applying environmental response functions.

Therefore, **mutual separability** of both types of effects is present if both (environmental and genotypic) effects are consistent and therefore separable from one another. However, **one-sided separability** can occur in two ways: separability of the genotypic but not environmental effects and vice versa.

Interaction means, that the effects are not consistent and therefore not separable. For example, if two genotypic response functions differ in one environment and do not so in another, or if the ranking of the response functions is not consistent, than the genotypic effects are said to interact with the environmneta1 effects and we call it **one-sided interaction**.

Mutual interaction occurs if both effects are not separable. Thus separability and interaction are complementary notions.

Results

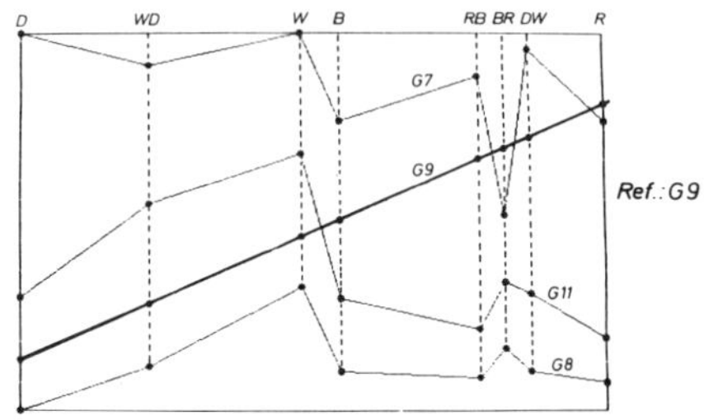

Fig 2: Callus growth; genotypic response function

The figures need some explanation, because the mode of demonstration is rather unusual. One of the genotypes is denoted as reference. The measurements for this genotyype are arranged with increasing values, so that the response function of this reference genotype is always a straight line. This fixes the environmental situations. The other genotypes are than plotted relative to this reference. It seems easier to compare the response functions with one another in this arrangement. Of course the same is true for the environmental response functions.

The reference genotype is G9. This graph of the norms of reaction demonstrates that G7, G11 and G8 are consistently distinguishable from one another over the whole sets of environments. However, G9 interacts with two genotypes (G7 and G11). Only G8 is separable from all genotypes. Here we have one sided separability of the genotypic from the environmental effects for G7, G8 and G11.

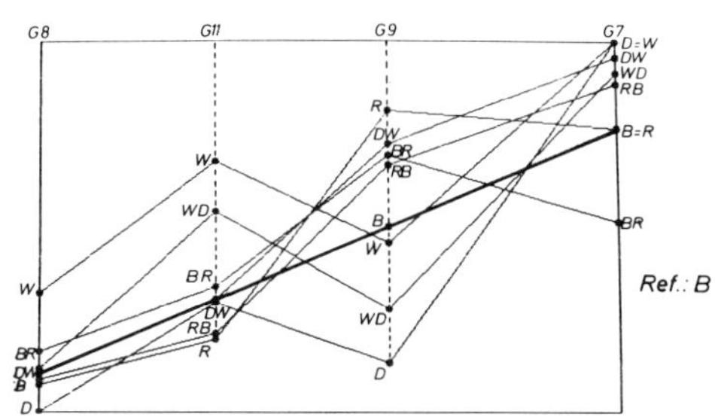

Fig. 3: Callus growth; environment response function

None of the environmental effects is consistent and interaction of the environmental with the genotypic effects takes place, but because of the separability of the genotypic effects (remember Fig. 2) the interaction is one-sided.

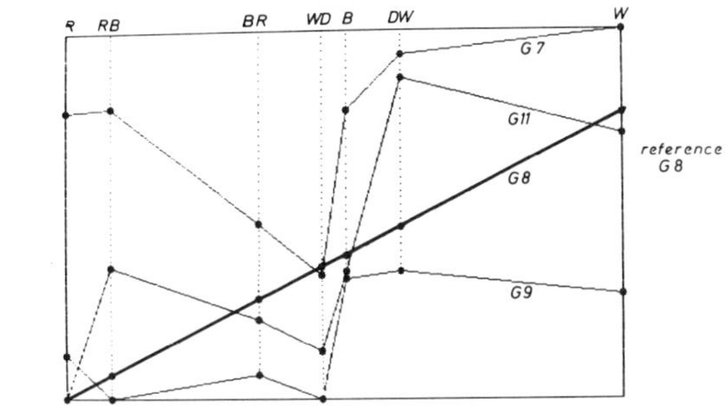

Fig. 4: Callus colour; genotypic response function

This figure demonstrates the other character. G7 is the genotype with the most green calli. Its reaction norm ranks above all others. Nevertheless, G7 is not consistently different to the other genotypes except to G9, so that interaction of the genotypic with the environmental effects take place and we can not differentiate between both effects.

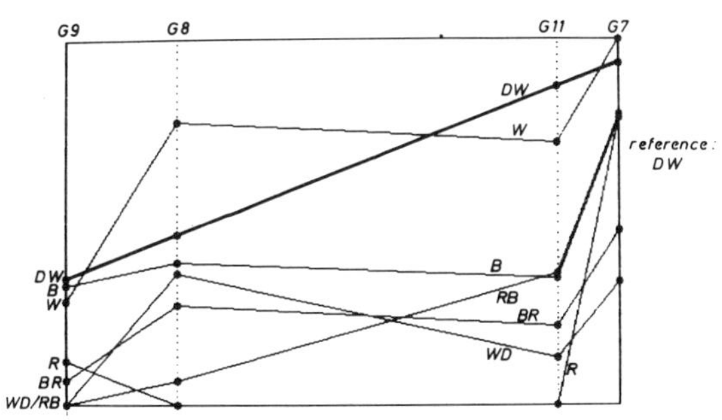

Fig. 5: Callus colour; environmental response function

Only two pairs of environmental situations can be termed as beeing separable: DW and RB on the one hand and W and BR (WD) on the other. But over all the environmental response functions are interacting. It is not possible to discriminate between the environmental and the genotypic effects in forming the phenotype.

So this is an example for mutual interaction of both effects, while the character callus growth is a case of one-sided interaction or separability.

Discussion

The most striking result, probably, is that the genotypic effects seem to be more consistent over the range of the applied environments. In other words, the genetic differences are stronger expressed at least in this experiment. The environmental effects are more likely suppressed by the genotypes in their effects on the phenotypes. Therefore it seems appropriate to relate the genetic differences to the different expression of the character growth and colour. Remember the figure to explain the different degree of kinship. Furtheron we have to notice both types of genetic information, the nuclear and the extranuclear.

Callus growth is thought to be under nuclear control. This is reflected in the results: The response functions of G7 and G8, the unrelated genotypes, are separated, while the norm of G9 resembles its halfsib G7 in some environments and its halfsib G11 in some others. G11, the genotype with the weakest relationship to G8 is separable, too.

With respect to callus colour the situation is more intricate. If it is true, that the genetic information of the chloroplast has an influence on the expression of this trait, than it could be expected that the varying light conditions would perform the greatest differences or similarities depend-dend on the degree of kinship of the seed parents, because birch is not known to belong to the group of species with biparental transmission of extranuclear genetic information. The results of the present experiment, however, suggest that the nuclear information may have the stronger influence even on this trait. But the observation that the response functions of G7 and G9 are clearly distinct can be explained by the fact that they are expected to differ in thier extranuclear genes because there is no relationship between their seed parents, while G8 and G9 are even likely identical in their extra-nuclear genes, since their seed parents are fullsibs.

It seems that the outcomes of this experiments may be important for tissue culturists, since experiments often concentrate on the environmental effects, mostly on media and the influence of the genetic variation is frequently ignored. However there may exist cases in which the behaviour of a culture depends to a larger degree on the selection of the genotypes than on the environmental conditions.

The application of the concept to experimental results has shown that this model of separability is very suitable to analyse data in the sense of causes and their effects. Above all, this method is suitable to answer practical questions for breeding purposes. For instance, if the results of a breeding experiment are not separable over all applied environmental conditions, then the character is not consistent and the "success" depends on the environments. But if it is possible to specify parts of a reaction norm which are separable than one can define a range of culture conditions where the superior genotypic effect consistently exists and perhaps one can offer special cultural conditions for special breeding purposes.

Likewise, the examination of the environmental effect can become very useful, too. Supposed, a group of genotypes should be cultured in an optimal environment one had to look for the environmental response function, which ranks consistently above the other environmental functions for all genotypes of this particular group. In this case it is asked for one-sided separability of the environmental effects from the genotypic effects and it is indifferent, wether the genotypic effects are interacting or not. The only important thing is, that the "optimal" environmnet is consistently different from the others and ranking above them.

Literature

Murashige, T., Skoog, F. (1962): A revised medium for the rapid growth and bioassay with tobacco tissue culture Physiol. Plant, 15, 473-497

Gregorius, H.-R, Namkoong, G. (1986): Joint analysis of genotypic and environmental effects. Theor. Appl. Genet., 72, 413-422

DETERMINATION OF PLANTLET REGENERATION CAPACITY OF SELECTED ASPEN CLONES IN VITRO

M.R. AHUJA, D. KRUSCHE and G.H. MELCHIOR

Federal Research Centre for Forestry and Forest Products, Institute of Forest Genetics and Forest Tree Breeding, Sieker Landstrasse 2, D-2070 Grosshansdorf, Federal Republic of Germany

0. ABSTRACT

European Aspen(*Populus tremula*), quaking aspen (*P. tremuloides*) and their hybrids are fast-growing, short rotation trees that have a wide range of adaptability and can be grown on marginal sites. Aspen wood has been employed in matchstick, paper- and plywood industries and has potential value for biomass and energy production. Vegetative propagation of aspen is extremely difficult by woody cuttings and uneconomical by root suckers and grafting. Consequently, large scale clonal multiplication of superior trees has not been feasible with conventional methods of vegetative propagation. On the other hand, tissue culture technology has been successfully exploited for large scale clonal propagation of aspen clones.

The present study was undertaken to determine the plant regeneration capacity and its association with growth performance, the genotype-culture media interaction and the optimal dose of 6-benzylaminopurine (BA) for tissues obtained from 15 clones of five aspen and hybrid aspen families.

It was found that there exists an optimal concentration of BA. The variation of microshoots per explant was significantly large for BA-treatments, families and clones within families. Certain families showed considerable genotype-culture media interaction. The regeneration capacity seems to be independent of the growth performance under field conditions, which is advantageous for breeding purposes.

1. INTRODUCTION

Aspen belong to the genus *Populus*, section Leuce, subsection Trepidae, of the family Salicaceae. European aspen (*Populus tremula*), quaking aspen (*Populus tremuloides*) and their hybrids are fast-growing with a wide range of adaptability and can be grown on marginal sites (25,29). Because of rapid growth, aspens appear to be potentially valuable for biomass and energy production (6). Aspen wood (27) has some peculiar characters suitable for many uses, including paper, plywood, and matchstick industries. Young aspen plants have also been used for production of animal feed. Antimicrobial substances have been detected in the bark and tissues of aspen (19). Because of these attributes, aspen are acquiring growing importance as forest trees and hybrid aspen breeding holds a high rank in our institutes breeding activities (20,23).

M. R. Ahuja (ed.), Somatic Cell Genetics of Woody Plants, 127–135.

Aspen have been mainly propagated by seed and families were introduced into practice according to the requirements of the law on forest reproductive material (20,21,22,25). To a limited extent propagation was done by root suckers, bud grafts, and green shoots (10,17,24). However these procedures are expensive. Vegetative propagation by woody cuttings is extremely difficult in aspen, because of lack of preformed root primordia on the stem cuttings. By employing conventional methods of vegetative propagation, it has not been possible to clonally propagate superior trees of aspen on a large scale. On the other hand, tissue culture technology has been succesfully exploited for large scale clonal propagation of aspen clones (1,2,3,9). The legal procedure of the transfer of new hybrid aspen varieties as clones and clonal mixtures into practice raises an other problem: the approval of a clone or a clonal mixture.

The present study was undertaken to determine the plant regeneration capacity and genotype-environment interactions of tissues obtained from 15 selected aspen and hybrid aspen trees.

2. MATERIAL AND METHODS

2.1. Culture of bud meristems

Dormant buds of 15 selected trees of European aspen (*Populus tremula*) and hybrid aspen (*P. tremula* x *P. tremuloides*) of 5 different families of an age of 13 years (see Table 1) were collected in October 1985 and stored at 0°C to 2°C for one week before culturing. Bud explants were prepared for *in vitro* culture according to the method described by Ahuja (3). Essentially bud meristems, with one or two young leaves, were cultured on a modified Woody Plant Medium (18), previously designated as Aspen Culture Medium (1). Four concentrations of 6-benzylaminopurine (BA), that is 0.2 mg l^{-1}, 0.5 mg l^{-1}, 1.0 mg l^{-1}, and 2.0 mg l^{-1} along with 0.02 mg l^{-1} of naphtaleneacetic acid (NAA) were tested for their effect on microshoot (MS) induction and proliferation. At least 10 bud explants were tested per treatment and genotype. The cultural conditions and rooting of microshoots (MS) in a soil-free potting mixture have already been described by Ahuja (2,3).

2.2. Statistical analysis

The following questions were investigated particularly:

- the optimal concentration of BA for microshoot induction
- family differences
- BA-treatment differences
- the differential response of families to treatments
- the correlation of MS induction with other morphological and physiological characters

The statistical analysis is based on the number of microshoots per explant (MS). MS was assumed as Poisson distributed with non-constant variance. Accordingly a log-linear regression model for MS was set up and was analysed with the statistical software package GLIM (8). Log-linear models take into account the heteroscedasticity of the data and use the maximum of the inherent information. The limited space allowed us not to present the complete analysis of deviance for each separate model, instead we communicate the results of the significance tests only.

TABLE 1. Origin of the clones investigated and number of microshoots per explant in clones and families.

family no.	origin	clone	no. of microshoots/explant: clone	no. of microshoots/explant: family
1	Grafted clone of a single tree P.t'a* from Tapiau/Ostpreussen, plantation near Wedesbüttel/Braunschweig	W 97	1.05	1.05
2	Controlled cross 169/59: P.t'a (W 3, Ostpreussen) x P.t's (Clone 13, Northern Wisc.)	KH 83	1.03	2.40
		KH 73		
3	Controlled cross 184/59: P.t'a (W 5, Ostpreussen) x P.t's (T 13-58, Middle Wisc.)	164 A	0.95	0.95
4	Controlled cross 19/51: P.t'a (Brauna 11, Sachsen) x P.t's (Turesson 141, New Hampsh.)	Esch 2	1.06	1.22
		Esch 5	0.6	
		Esch 6	1.5	
		Esch 8	1.5	
		Se 1	0.7	
		Se 2	1.75	
		Se 3	1.4	
5	Controlled cross 8/53: P.t'a (Großdubrau 1, Sachsen) x P.t's (Maple, Ontario)	Wä 1	3.97	2.2
		Wä 8	1.62	
		Wä 13	1.7	
		Wä 14	1.65	

* P.t'a = *P. tremula*, P.t's = *P. tremuloides*

3. RESULTS

The number of microshoots per explant (MS) varied between 0 and 10, with the most frequent MS of 1 and 2.

3.1. Response of MS to BA-treatments

The overall response of MS to the four BA-treatments is depicted in figure 1. The differences between BA-levels are highly significant, with a maximal response of MS for 0.5 mg l^{-1} BA. A quadratic response function described the data accurately, with an average of 2.3 microshoots/explant at the maximum.

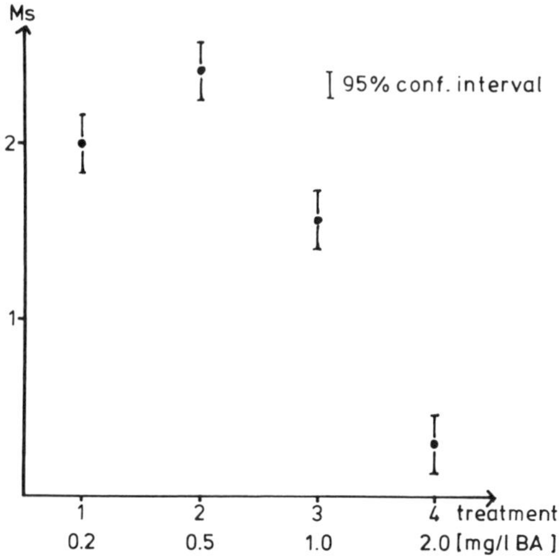

FIGURE 1. The number of microshoots per explant (MS) averaged over all explants of all clones for each BA-treatment.

3.2. Response of MS to families and BA-treatments

We found large significant differences for MS production between families as well as between BA-treatments (figure 2). Families 1, 4 and 5 varied in a similar manner, with a maximum of MS at the 0.5 m l^{-1} level of BA. The response of family 2 was nearly constant for MS equal to 2.5, whereas family 3 showed a steady decline of MS with no apparent maximum. This differential response of families to BA-treatments accounts for the statistical significant family by treatment interaction component.

3.3. Correlation of MS induction with other morphological and physiological characters

In table 2 the rank correlation coefficients between MS and some other important characters of the investigated clones are given. The rootability refers to the percentage of successfully rooted microshoots in a peat-perlite substrate, whereas the microshoot harvest index is a measure of the time taken by an explant to be ready for cutting the first set of microshoots for rooting in the substrate. The main result is that the growth performance and volume production of the clones is not correlated to the ability for MS proliferation.

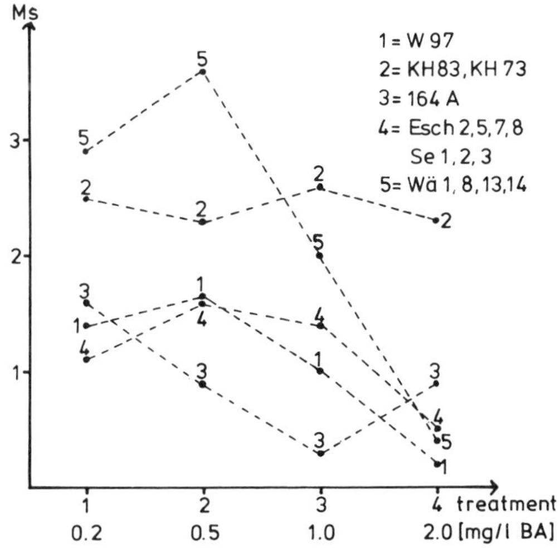

FIGURE 2. The number of microshoots per explant (MS) averaged over all explants of all clones of each family and for each BA-treatment.

Furthermore the rootability of a clone tends to be highest (r=.41) under BA-level 0.5 mg l^{-1}, at least there seems to rooting success in the substrate. A negative correlation between microshoot harvest index and MS-induction was observed, i.e. clones with a low MS-induction take a longer time until harvest of microshoots.

4. DISCUSSION

4.1. Optimal BA concentration for MS induction

The results of the present study on 15 aspen clones have revealed that the BA concentrations for MS induction lie between 0.2 to 1.0 mg l^{-1} range. Our previous investigations with other aspen clones (1,2) and those of other investigators working with aspen (13,28,30) have also shown that the middle value (0.5 mg l^{-1}) may be the optimal BA concentration for induction and proliferation of MS in vitro. By employing 0.5 mg l^{-1} BA along with a very low concentration of NAA (0.02 mg l^{-1}) on a modified Woody Plant Medium, almost direct differentiation of MS was obtained on the bud meristem. There was hardly any visible callus formation on the bud explants before microshoot induction or proliferation. Microshoots were rooted in a peat-perlite subtrate. By emploing this two-step micropropagation method (induction and proliferation of MS *in vitro*, and rooting of MS *in vivo*) (2), it has been possible to minimize the exposure of tissues to unnatural *in vitro* conditions. Extensive callus formation in long-term cultures may be accompanied by genetic variability under *in vitro* conditions (5). For mass cloning of selected genotypes, it is essential to minimize the culture-induced genetic variability so that true-to-type clones are produced.

TABLE 2. Rankcorrelation of the number of microshoots/explant (MS) for the four BA-treatments with stem volume, diameter and height growth, microshoot harvest index* and rootability on a clonal basis (n=15).

	MS T1	MS T2	MS T3	MS T4
MS T1	1			
MS T2	0.54	1		
MS T3	0.70	0.42	1	
MS T4	0.48	-0.18	0.48	1
volume 1986	0.35	0.36	0.08	-0.20
d.b.h. 1982	0.09	0.12	-0.09	-0.22
d.b.h. 1986	0.05	0.15	-0.22	-0.39
height 1986	0.53	0.34	0.46	0.07
rootability	0.36	0.41	0.22	-0.06
harvest index	-0.78	-0.40	-0.38	-0.30

* Time taken (in weeks) by a clone to be ready for the harvest of the first set of microshoots for rooting in a peat-perlite substrate (1 = 10 weeks, 2 = 14 weeks, 3 = 17 weeks).

4.2. Differential response of families to BA treatments

In terms of MS production on the bud explants, there were significant differences between BA treatments. In addition, there are also significant differences between families with respect to overall MS production. The overall response of family 2 and 5 was nearly twice as much as the response of the other families. Most of the families varied in a somewhat similar manner, with a peak of MS production at 0.5 mg l^{-1} BA. One family (family 2) remained nearly constant on all BA treatments, while the remaining family (family 3) showed a steady decline of MS production on all BA concentrations. So for most families an optimal BA concentration can be reached for maximum MS induction. The differential family-treatment response and extent of MS suggests that the number of MS production on the primary bud explants seems to be genetically controlled.

4.3. Correlations between MS and other traits

The correlation of MS induction with several other traits, including volume growth, diameter growth and MS harvest index were investigated on a clone basis. The rank correlation analysis (Table 2.) suggested that the MS induction potential of a clone is not correlated with the volume and growth performance. In other words, there is no correlated selection against growth and volume of a tree when selection is carried out for MS induction on its bud explants. We found a negative correlation between MS inducibility and MS harvest index, that is, the clones with a low MS induction take a longer time until the microshoots can be harvested for rooting. These results indicate that the regeneration capacity seems to be independent of growth performance of a clone under field conditions, which has advantageous implications for breeding purposes and the establishment of clonal mixtures.

4.4. Aging and micropropagation

Clonal propagation of most mature hardwood and softwood tree species are generally difficult by tissue culture. However, some progress has been made in this direction. By culture of bud explants, clonal propagation has been achieved in 100 years old *Tectona grandis* (15), 60 years old *Prunus avium* 4), 20 years old *Eucalyptus citriodora* (16), and 17 to 40 years old aspen (*Populus tremula* ,*P. tremuloides*) and hybrid aspen (*Populus tremula* x *P. tremuloides*) (1,2,3,4). Micropropagation of up to 100 years old *Sequoia sempervirens* (7,12) and 25 to 30 years old *Larix decidua* (11) have also been reported.

In previous studies with aspen, Ahuja (3,4) employed bud explants from more than 100 mature trees that were between 17 to 40 years of age. In these studies approximately 50 percent of the clones were responsive to *in vitro* micropropagation. In the present study, bud explants from all the 13 years old trees responded to the *in vitro* conditions by production of microshoots. There seems to be no aging effect on micropropagation in the 15 aspen clones employed in this study. There could be at least two explanations for this success: (i) the clones are probably not old enough to elicit an aging reponse; and (ii) the number of clones is too small and it is conceivable that by chance all the 15 aspen clones responded to the spectrum of broad cultural conditions. Although increasing age may reduce the *in vitro* growth and differentiation potential of an aspen tree, it is probable that the growth response, to a large extent, still remains under genetic control.

4.5. Approval and release of clones into practice

The official procedure for the transfer of new varieties, clones and clonal mixtures of Federal Research Organisations into practice (21) raises some problems. The approval of the material only takes place after the evaluation of a clonal field test generally at an age of 5 to 10 years. The material must be superior in growth performance and other desirable traits, when compared to a given standard. Unfortunately the vegetative propagation rate decreases with increasing age. The propagation rate is also substantially determined by the genotype. Therefore, the age and the genotype of the clones may impose certain limitations on the regenerative potential. This situation had to be overcome, when the approval by the designated authority was given for the clonal mixture called "Großhansdorf" with 14 hybrid aspen clones of unknown propagation rate. As the official transfer into practice required the utilization of all clones in the 14 clones mixture, we had to find a suitable micropropagation method for all clones. This was indeed accomplished by the present study employing a cost-effective, commercially feasible two-step micropropagation method.

REFERENCES

1. Ahuja MR: Somatic cell differentiation and rapid clonal propagation of aspen. Silvae Genet. 32, 131-135, 1983.
2. Ahuja MR: A commercially feasible micropropagation method for aspen. Silvae Genet. 33, 174-176, 1984.
3. Ahuja MR: Aspen. In: Handbook of Plant Cell Culture, Vol 4, Evans DA, Sharp WR and Ammirato PJ, (Eds.), Macmillan Publishing Company, New York, pp. 626-651, 1986.
4. Ahuja MR: In vitro propagation of Poplar and Aspen. In: Cell and Tissue Culture in Forestry, Vol 3, Case Histories: Gymnosperms, Angiosperms and Palms, Bonga JM and Durzan DJ, (Eds.), Martinus Nijhoff Publishers, Dordrecht, pp. 207-223, 1987.
5. Ahuja MR: Somaclonal variation. In: cell and Tissue Culture in Forestry, Vol 1, General Principles and Biotechnology, Bonga JM and Durzan DJ, (Eds.), Martinus Nijhoff Publishers, Dordrecht, pp. 272-285, 1987.
6. Ahuja MR and Muhs HJ: Biomasseproduktion in forstlichen Plantagen für die Rohstoff- und Energiegewinnung. Bundesministerium für Forschung und Technologie, Forschungsbericht T-86-073, Bonn, pp. 1-38, 1986.
7. Ball EA, Morris DM and Reydelius JA: Cloning of Sequoia sempervirens from mature trees by tissue culture. In: In Vitro Manipulation of Woody Plants, Boxus P, (Ed.), Cen - re de Recherches Agronomiques de l'Etat Ministere de l'Agriculture, Gembloux, pp. 181-226, 1978.
8. Baker RJ and Nelder JA: The Glim System, Release 3. National Algorithm Group, Oxford, 1978.
9. Barocka HK, Baus M, Lontke E and Sievert F: Tissue Culture as a tool for in vitro mass Propagation of aspen. Z. Pflanzenzüchtung 94, 340-343, 1985.
10. Behrens V und Melchior GH: Hinweise zur Verbesserung des Verfahrens der vegetativen Vermehrung von Aspen und Aspenhybriden durch Grünstecklinge aus jungen Wurzelschößlingen. Holzzucht 32, 17-21, 1978.
11. Bonga JM: Attempts to micropropagate mature Larix decidua Mill. This Volume.
12. Boulay M: Multiplication et clonage rapide on Sequoia sempervirens par la culture in vitro. In: Micropropagation d'Arbres Forestiers, Ann AFOCEL, Nangis, No 12, 6/79, pp. 49-56, 1979.
13. Christie CB: Rapid propagation of Aspen and Silver poplar using tissue culture techniques. Proc. Int. Plant Prop. Soc. 28, 255-260, 1978.
14. Cornu C and Chaix C: Multiplication par culture in vitro de merisiers adultes (Prunus avium)- application a un large eventail de Clones. In: Proc IUFRO Sect. S2.01.05, Int. Workshop "In Vitro Cultivation for Tree species ", Fontainebleau, pp. 71-79, 1981.
15. Gupta PK, Nadgir AL, Mascarehnas AF and Jagannathan V: Tissue culture of forest trees : clonal multiplication of Tectona grandis L. (teak) by tissue culture. Plant Sci Lett 17, 259-268, 1980.

16.Gupta PK, Mascarehnas AF and Jagannthan V: Tissue culture of forest trees-clonal propagation of mature trees of Eucalyptus citridora Hook. by tissue Culture. Plant Sci Lett 20, 195-201, 1981.
17.Herrmann S und Seuthe H: Vegetative Vermehrung von Hybridaspen durch Okulation. Holzzucht 36, 17-22. 1982.
18.Lloyd G and McCown BH: Commercially feasible micropropagation of mountain laurel (Kalmia latiflora) by use of Shoot tip culture. Proc Int Plant Prop Soc 30, 421-427, 1981.
19.Mathes MC: Antimicrobial Substances from aspen tissue grown in vitro. Science 140, 1101-1102, 1963.
20.Melchior GH: Züchtung von Aspen und Hybridaspen und ihre Perspektiven für die Praxis. Allgem. Forst- und Jagdzeitung 156, 112-122, 1985.
21.Melchior GH: Procedure to transfer breeding results of federal research organizations into practical use by administrave means. Mitt. Bundesforschungsanstalt für Forst- und wirtschaft, Hamburg. No. 154, 125-133, 1986.
22.Melchior GH, Krusche D and Sachsse H: Wachstum, holztechnologische Eigenschaften und züchterische Aspekte der Hybridaspenfamilie "Holsatia". Holzzucht 41, 25-29, 1987.
23.Melchior GH und Seitz FW: Einige Ergebnisse bei Testanbauten mit Aspenhybriden. 1. Kreuzungen des Jahres 1951. Silvae Genetica 15, 127-133, 1966.
24.Melchior GH und Seuthe H: Veredelungsunverträglichkeit zwischen Aspenarten und Aspen- und Hybridaspenunterlagen. Gartenbauwissenschaft 50, 170-171, 1985.
25.Mohrdiek O: Hybridaspen für forstliche Grenzertragsböden. Forstarchiv 48, 158-163, 1977.
26.Muhs HJ: A synopsis of some international and national rules for forest reproductive material moving into trade-OECD-Scheme, EEC Directives and German Law. of forest reproductive material. Mitt. Bundesforschungsanstalt für Forst- und Holzwirtschaft, Hamburg. No 154, 25-37, 1986.
27.Sachsse H, Supraptono B und Melchior GH: Über das Holz der Hybridaspenfamilie 19/51 ("Holsatia") aus zwei Versuchsorten. Holzforschung 38, 297-308, 1985.
28.Wann SR and Einspahr DE: Reliable plantlet formation from seedling explants of Populus tremuloides (Mich.). Silvae Genet 35, 19-24, 1986.
29.Weisgerber H: Untersuchungen über das Wuchsverhalten von Aspenhybriden auf ungünstigen Waldstandorten in Deutschland. Holzzucht 30, 48-53, 1976.
30.Wolter KE: Root and shoot initiation in aspen callus cultures. Nature 219, 509-510, 1968.

SUSPENSION CULTURE OF THE DIPTEROCARP *SHOREA ROXBURGHII* G. DON.

D.S. GUNASEKARA, E.S. SCOTT AND A.N. RAO
Botany Department, National University of Singapore, Lower Kent Ridge Road, Singapore 0511, Republic of Singapore

ABSTRACT

Liquid cultures of very few tree species have been grown *in vitro* to obtain cell suspensions as a prelude to somatic embryogenesis. The few reports published on *Ulmus*, *Liquidambar* and *Santalum* species indicate that plantlets can be obtained either through direct embryos or from nodule-like structures that develop into shoots. The present paper reports the establishment of suspension cultures from *Shorea roxburghii* (Dipterocarpaceae) cotyledon callus and attempts to induce organogenesis. The dipterocarps are the major timber trees of S.E. Asia with very high export value.

1. INTRODUCTION

The *Dipterocarpaceae* comprises many tropical hardwood species and members of the genus *Shorea* are a particularly valuable source of timber in S.E. Asia. Due to heavy logging, dipterocarp forest is being depleted and natural regeneration of some species in logged forest is slow and limited (5,12). Phenological data on dipterocarps are scarce. Tissue culture may facilitate the production of plantlets for reforestation, genetic improvement and conservation of germplasm of these valuable timber trees.

There have been few reports of tissue culture of dipterocarps. Scott *et al.* (9,10) produced plantlets of *Shorea roxburghii* G. Don. and *Hopea odorata* Roxb. by rooting axillary shoots which had been induced from embryonic axes and embryos respectively. Cotyledons of both the above species produced callus and/or roots in the presence of 2,4-D. Smits and Struycken (12) likewise induced callus and roots in leaf explants of *S. curtisii*. No attempts at suspension culture have been reported to date.

Relatively few tree species have been grown in cell suspension form. Durzan and Lopushanski (3) established suspension cultures from hypocotyl callus of American elm (*Ulmus americana*) which, when plated on agar medium without growth regulators, gave rise to red and green nodules which later developed into shoots. Sommer and Brown (13) grew hypocotyl callus of American sweetgum (*Liquidambar styraciflua*) in liquid medium and after 4 months embryoids capable of developing into normal plants were obtained. Embryogenesis was also reported in suspension cultures of sandalwood (*Santalum album*) (11).

This paper reports the establishment of suspension cultures from *Shorea roxburghii* cotyledon callus, subsequent growth studies, and attempts to induce organogenesis.

2. MATERIALS AND METHODS

2.1. *Cell culture*. Callus tissues were derived from cotyledons of *Shorea roxburghii* G. Don. cultured on a modified MS (8) medium (9) containing 4.5 x 10^{-6} M 2,4-dichorophenoxyacetic acid (2,4-D) and 4.4 x 10^{-7} M

M. R. Ahuja (ed.), Somatic Cell Genetics of Woody Plants, 137–141.

6-benzylaminopurine (BA) and were maintained, by monthly subculture on the same medium, for 1 year.

The modified MS formulation was used as basal medium for suspension culture, supplemented with 3% (w/v) sucrose unless otherwise stated and pH adjusted to 5.6. Cell suspension cultures were initiated by placing 0.8-1.0 g callus in 20 ml liquid medium in 100 ml conical flasks and agitating on a rotary shaker at 100 rpm at 27 ± 2°C in darkness or low light intensity (c. 2.5 $\mu Em^{-2}s^{-1}$). The effects of additives such as coconut water (CW: 5% v/v), casein hydrolysate (CH: 0.01% w/v) and the growth regulators 2,4-D (0-4.5 x 10^{-5} M) and BA (0-4.4 x 10^{-6} M) on growth of cell suspensions were studied. The effect of variation in sucrose concentration (3-18%) was also investigated.

2.2. Stock cultures. Primary cultures showing most vigorous growth were maintained as stock cultures (Fig. 1a), in MS with CW (5% v/v), 4.5 x 10^{-7} M 2,4-D and 4.4 x 10^{-8} M BA, for use in subsequent experiments; 1 ml stock suspension was used to inoculate 20 ml medium.

2.3. Growth curves. Cells were counted at 4 day intervals using a haemocytometer (Neubauer type); 30 counts were made for each treatment.

2.4. Plating on solid medium. One ml of 1 month-old stock suspension was diluted 10 times and spread evenly over agar medium in 9 cm petri dishes or was mixed with 9 ml molten agar and poured into petri dishes to give a 1 mm layer. The agar medium was supplemented with 0, 4.5 x 10^{-7}, 4.5 x 10^{-6}, 2.2 x 10^{-5} M 2,4-D plus 4.4 x 10^{-8} or 4.4 x 10^{-7} M BA. Nurse calli were placed on some replicates and all cultures were kept in the dark. Clumps of callus and roots formed in suspension cultures were also plated on agar medium supplemented with various combinations of BA, kinetin (K), zeatin (Z), indolebutyric acid (IBA) and naphthaleneacetic acid (NAA).

2.5. Observations. Cultures were examined using an inverted microscope and cell samples were stained with lactophenol cotton blue or aceto-orcein to facilitate morphological studies.

3. RESULTS AND DISCUSSION

3.1. Establishment and growth of suspension cultures. The addition of CW and/or CH greatly enhanced the growth of cells in suspension, for all growth regulator combinations tested. Clumps of cells in basal liquid medium, with or without CW or CH, developed roots after 1-2 weeks (Fig. 1b). These roots elongated and eventually formed lateral roots.

The growth curves of cells in medium with CW and various concentrations of 2,4-D are shown in Fig. 2. 2,4-D at 4.5 x 10^{-7} and 4.5 x 10^{-6} M stimulated growth, with mean values of 2.55 and 2.53 x 10^{6} cells per ml respectively after 32 days; growth in medium with 0, 2.2 x 10^{-5}, 4.5 x 10^{-5} M 2,4-D was poor. Growth curves of cells in medium supplemented with 2,4-D and BA were erratic and cell numbers were apparently lower. This phenomenon was due to proliferation of cell clumps in the presence of BA, thus cell numbers were under-estimated. This problem may be overcome by using chromic acid, pectinase or cellulase to disperse cell aggregates (2). Nevertheless, cell numbers in medium with 4.5 x 10^{-7} M 2,4-D and 4.4 x 10^{-8} M BA were significantly higher than in other treatments of this series. Thus, when relatively homogeneous suspensions of single

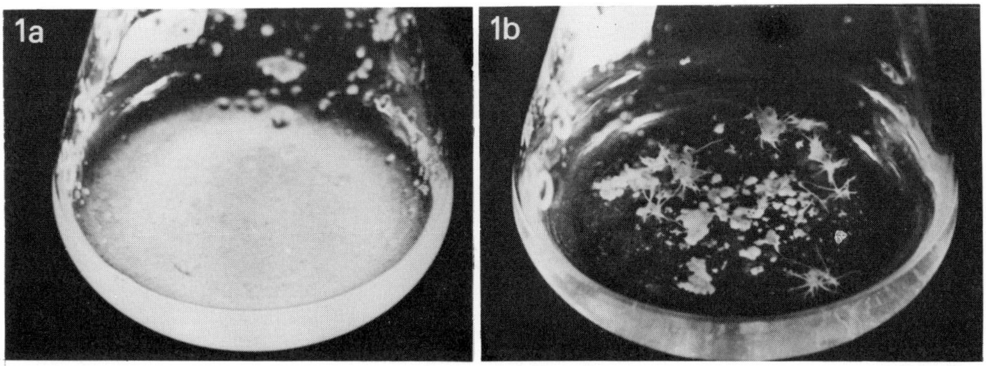

Figure 1. Cell suspension of Shorea roxburghii cultured for 4 weeks in a) modified MS medium with 5% (v/v) coconut water, 4.5×10^{-7} M 2,4-D and 4.4×10^{-8} M BA. b) basal medium, showing development of roots.

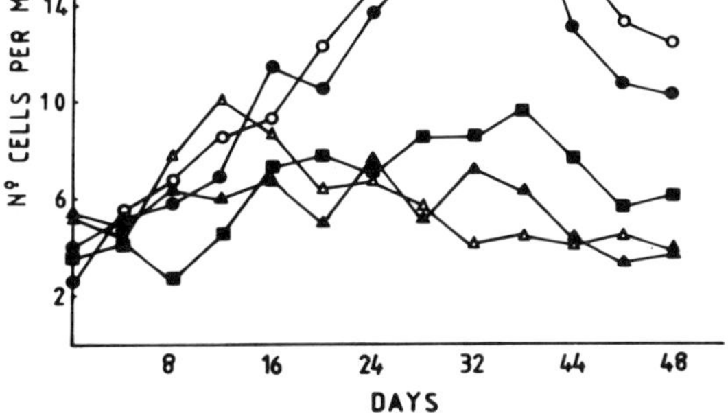

Figure 2. Growth curves of suspension cultures of Shorea roxburghii in modified MS with coconut water and 2,4-D, ■: 0, ○: 4.5×10^{-7}, ●: 4.5×10^{-6}, ▲: 2.2×10^{-5}, △: 4.5×10^{-5}, M.

cells are required, 4.5 x 10^{-7} M 2,4-D alone may be used whereas 4.4 x 10^{-8} M BA may be added to the medium if large amounts of material, regardless of homogeneity, are needed.

3.2. Plating suspensions on agar medium. Single cells failed to form colonies under the conditions employed, whereas small clusters of cells did divide to form clumps of callus. Nurse calli had little effect except on basal medium, where they enhanced growth of cell clusters. It is possible that single cells were damaged during isolation or that they were unable to adapt to the phase change from liquid to agar medium where nutrients are less readily available and inhibitory metabolites may build up around the cells (1,14). Most studies of this type have involved non-woody plants; little work has been done with single cell culture of woody plants.

Clumps of callus, and roots formed in basal liquid medium, were plated on agar medium supplemented with various combinations of growth regulators and incubated under cool daylight fluorescent lamps (25 $\mu Em^{-2}s^{-1}$), 14 h day-length. Roots plated on basal agar medium elongated and formed lateral roots whereas those on media with cytokinins formed callus; the root epidermis split and beige coloured callus emerged. This callus did not develop further. Callus plated on basal medium and on medium with low concentrations of auxins and cytokinins gave rise to roots after 3 weeks. Little growth was observed in medium with BA or K alone or in combination with IBA. Callus on medium with 2.2 x 10^{-6} M BA and 2.3 x 10^{-6} M K, however grew well and turned green but failed to differentiate. Further experiments to induce organogenesis, particularly shoot development, are underway.

3.3. Observations on cells in suspension. Cultures in liquid medium with 3-9% sucrose grew well but there was no evidence of organogenesis or embryo formation. Increasing the concentration of sucrose has been shown to induce embryogenesis in various species including soybean (*Glycine max*) (6) but had no apparent effect on *S. roxburghii*.

Staining and microscopic examination revealed small, densely cytoplasmic cells forming small clusters in liquid medium supplemented with CW, CH and various combinations of 2,4-D and BA. Such cells are capable of division and are potentially embryogenic (4,7). Many 3-4-cell clusters were observed in suspension cultures of *S. roxburghii*; while such clusters have been shown to be proembryonic masses in other species, there was no conclusive evidence of further embryogenesis in this case.

4. CONCLUSIONS

This study has shown that dipterocarp cells can be grown readily in suspension culture, thus large amounts of material can be produced for further studies such as induction of organogenesis, embryogenesis and protoplast culture. Limited organogenesis, in the form of root production has been obtained and experiments to induce shoot formation and embryogenesis, are now in progress.

ACKNOWLEDGEMENTS

The work was supported by the National University of Singapore (research grants RP120/84 and GP006/85).

REFERENCES

1. Bergman L: Growth and Division of Single Cells of Higher Plants *In Vitro*. Journal of General Physiology, 43, 841-851, 1960.

2. Bhojwani SS and Razdan MR (eds): Plant Tissue Culture Theory and Practice. Elsevier, New York, 1983.
3. Durzan DJ and Lopushanski SM: Propagation of American Elm via Cell Suspension Culture. Canadian Journal of Forest Research, 5, 273-277, 1975.
4. Halperin W: Alternative Morphogenetic Events in Cell Suspensions. American Journal of Botany, 53, 433-453, 1966.
5. Jacobs M: Dipterocarpaceae: the Taxonomic and Distributional Framework. Malaysian Forester, 44, 168-189, 1981.
6. Kerns HR, Barwale UB, Meyer MM and Widholm JM: Correlation of Cotyledonary Node, Shoot Proliferation and Somatic Embryoid Development in Suspension Cultures of Soybean (Glycine max L. Merr.). Plant Cell Reports, 5, 140-143, 1986.
7. McWilliam AA, Smith SM and Street HE: The Origin and Development of Embryoids in Suspension Cultures of Carrot (Daucus carota). Annals of Botany, 38, 243-250, 1974.
8. Murashige T and Skoog F: A Revised Medium for Rapid Growth and Bioassays with Tobacco Tissue Cultures. Physiologia Plantarum, 15, 473-479, 1962.
9. Scott ES, Rao AN and Loh CS: Production of Plantlets of Shorea roxburghii G. Don. from Embryonic Axes Cultured In Vitro. In Press.
10. Scott ES, Rao AN and Loh CS: Tissue Culture of Hopea odorata Roxb., a Dipterocarp Tree. In Press.
11. Sita GL, Raghava Ram NV and Vaidyanathan CS: Triploid Plants from Endosperm Cultures of Sandalwood by Experimental Embryogenesis. Plant Science Letters, 20, 63-69, 1980.
12. Smits WThM and Struycken B: Some Preliminary Results of Experiments with In Vitro Culture of Dipterocarps. Netherlands Journal of Agricultural Science, 31, 233-238, 1983.
13. Sommer HE and Brown CL: Embryogenesis in Tissue Culture of Sweetgum. Forest Science, 2, 257-260, 1980.
14. Torrey JG: Cell Division in Isolated Single Plant Cells In Vitro. Proceedings of National Academy of Science (US), 43, 887-891, 1957.

ISOZYME ANALYSIS OF MICROPROPAGATED *BETULA PENDULA* ROTH.

GRZEGORZ KOSINSKI, AMANDA M GORDON AND IAN R BROWN
DEPARTMENT OF FORESTRY, UNIVERSITY OF ABERDEEN, ST MACHAR DRIVE, ABERDEEN AB9 2UU, SCOTLAND

INTRODUCTION

Micropropagation is becoming an increasingly popular technique used by forest tree breeders to circumvent the problems of long life cycle, prolonged juvenility and disturbance in flower and seed production. To date, however, little is known concerning the inherent genetic changes which may accompany the micropropagation process.

Tissue culture of birch has been reported by a number of workers (Huhtinen and Yahgaoglu, 1974; McCown and Amos, 1979; Srivastava and Steinhauer, 1981a, b; Srivastava *et al*., 1985). These studies indicated that micropropagated plants exhibit high growth uniformity, lack of gross chromosomal changes and preservation of morphological identity and early flowering characteristics. Srivastava and Steinhauer (1981a) report that an alteration in isozyme expression accompanies organogenesis in birch callus. This is not entirely unexpected as the material is proceeding through a developmental change with accompanying alterations in gene expression. At present there is little experimental evidence to indicate whether clonal micropropagated plantlets are genetically identical individuals. These studies have been undertaken to determine the level of variation which occurs during the micropropagation of *Betula pendula*.

PROCEDURE

Materials and Methods

Seeds of *Betula pendula* Roth. were sterilized and germinated *in vitro*. The roots were removed and the seedling was cut into two sections. The upper section was micropropagated by nodal cuttings in half strength Murashige and Skoog (1962) medium with 30% (w/v) sucrose and 0.6% (w/v) agar. The lower section was micropropagated in a similar media with the addition of 2mg l^{-1} N^6 - benzylamino purine (BAP).

For isozyme (esterase, Est and peroxidase, PX) determination shoots were removed at regular intervals. They were homogenized in tris-HCl (pH 8.0) buffer containing 10% (w/v) insoluble polyvinylpyrrolidine. The extracts were analysed by discontinuous vertical polyacrylamide gel-electrophoresis (LKB Bromma). Proteins were separated on 12% (w/v) acrylamide gels, pH 8.3 at 25°C with a current of 60 mA.

RESULTS

For ten *Betula pendula* Roth. clones peroxidase activity was exhibited by a monomorphic phenotype consisting of two bands (Fig 1). Seven different multi-banded phenotypes of esterase were observed (Fig 1). Two of the *Betula pendula* clones, G5 and G9, exhibited differing esterase patterns depending on the method of micropropagation ie. whether the material was derived from the upper or lower sections of the seedling. Phenotype 1 represents the pattern obtained for nodal cuttings from the upper half of a G5 seedling and phenotype 2 that from plantlets multiplied

M. R. Ahuja (ed.), Somatic Cell Genetics of Woody Plants, 143–145.

by shoot proliferation in the presence of BAP. The corresponding plantlets of G9 were characterized by phenotypes 3 and 4 respectively. For all clones analysed there was no evidence of any variation between the individuals multiplied by a similar method of micropropagation. The isozyme phenotypes characteristic of these clones remained unchanged independent of the length of time in culture.

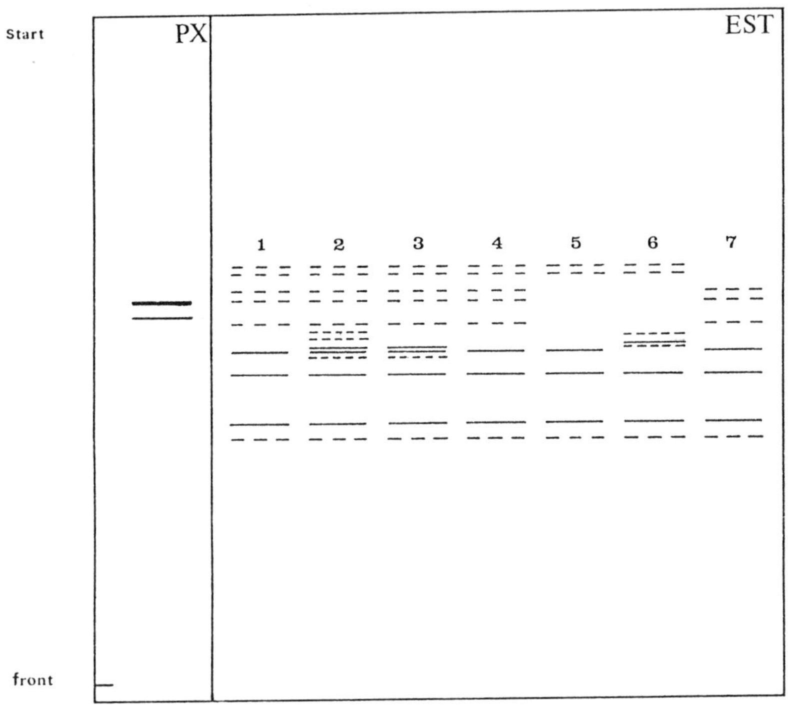

FIGURE 1. Esterase (EST) and peroxidase (PX) isoenzymes. Diagrammatic comparison of isoenzymes from a number of Betula pendula Roth. clones. The clones of interest G5 and G9 are represented by columns 1 and 3 for nodal cuttings and columns 2 and 4 for proliferated shoots. The density of the bands gives an indication of the level of activity.

DISCUSSION

During the last decade a number of methods of micropropagation have been utilised to allow regeneration and clonal multiplication of selected plant material. Most of these involve the inclusion of cytokinins in the media to encourage shoot proliferation. The technique described is used routinely in our laboratory and is a successful method for the micropropagation of both Betula pendula and Betula pubescens from seedlings.

To date, there is little experiemental evidence regarding the degree of variation which occurs as a result of genetic instability during the micropropagation of forest trees. As described in the introduction, a number of workers have reported that micropropagated birch trees appear to retain their original phenotypic characteristics.

Of the two isozyme systems, peroxidase and esterase, the latter proved to be potentially useful as a marker of genetic instability of micropropagated material. Two of the clones exhibited differing esterase patterns depending on the micropropagation technique. These changes in the shoot culture appear to be stable as successive generations of plantlets retained the altered banding pattern. Similarly there was no change in the isozyme patterns from plantlets grown for three months after removal from the "parent" culture. The factor or factors responsible for the altered isozyme patterns are as yet undetermined. The variation does not appear to be a consequence of the direct effect of BAP on gene expression. Nodal cuttings of G5 and G9 were grown for four weeks on medium with or without BAP and this did not result in any change in the esterase banding patterns. Moreover, the lack of intra-clonal differences between similarly micropropagated plantlets suggests that the observed changes are not due to somaclonal variation. Instead, it would appear that a universal change has taken place throughout the source explant which affects each individual in an identical manner. To our knowledge, this is the first report of this type of culture-induced variation.

REFERENCES

1. Huhtinen, O and Z Yahgaoglu: Das fruhe bluhen von aus Kalluskulturen herangezogen Pflanzchen bei der Birke (Betula pendula Roth). Silvae Genetica, 23, 32-34 (1974).
2. McCown, B and R Amos: Initial trials with commercial micropropagation of birch selections. Int. Plant Propag. Soc. Comb. Proc. 29, 387-393 (1979).
3. Murashige, T and F Skoog: A revised medium for rapid growth and bio-assays with tobacco tissue cultures. Physiol. Plant. 15, 473-497 (1962).
4. Srivastava, P S and A Steinhauer: Isozymes in differentiating shoot bud cultures of Betula pendula Roth. Z. Phlanzenphysiol. 103, 341-346 (1981a).
5. - - Regeneration of birch plants from catkin tissue cultures. Plant Sci. Lett. 22, 379-386 (1981b).
6. Srivastava, P S; A Steinhauer and H Glock: Plantlet differentiation in leaf and root cultures of birch (Betula pendula Roth.). Plant Sci. 42, 209-214 (1985).

ACKNOWLEDGEMENTS

This work was supported by grants from the Natural Environment Research Council and the British Council.

TISSUE CULTURE STUDIES ON MAHOGANY TREE, *Sweitenia*

S. VENKETESWARAN, M.A.D.L. DIAS, F. SULTANBAWA, U.V. WEYERS
Dept. of Biology,, University of Houston, Houston, Texas 77004. U.S.A

1. ABSTRACT

The mahogany tree, *Sweitenia,* which grows as natural populations in tropical America and parts of Asia and Africa, is a highly valued timber tree known for its redwood color, strength, durability, water resistance and aesthetic appeal. Callus tissues were established from aseptic segments of cotyledons and leaves of mahogany (*Sweitenia mahogany* Jacquin and *S. macrophylla* King). They were maintained on a modified Murashige & Skoog (MS) medium supplemented with 2 mg/l of 2,4-dichlorophenoxyacetic acid (2,4-D) and 1 mg/l of -naphthalene acetic acid (NAA). The explants proliferated extensively on agar medium in the growth chamber at $25 \pm 1^\circ C$ in 2-3 weeks. The callus from cotyledonary tissues were very soft and white in color and further subcultures resulted in a loose "snow-flake" appearance. Callus obtained from leaves were almost transparent and colorless. Experiments on *in vitro* culture and micropropagation of mahogany suggest that mass propagation of this tree is possible using this method, particularly because of decline in natural populations.

2. INTRODUCTION

The Mahogany tree, *Sweitenia* is a large tropical American tree growing over 100 ft high and reaching around 4-5 ft in diameter. The heartwood of this tree is well-known and is considered as a highly-valued timber because of its strength, durability and water resistance. The wood is reddish, pinkish, salmon-colored or yellowish when fresh and deepens with age to a reddish brown golden luster. Because of these qualities, the wood is used in the manufacture of furniture, interior decoration, boat construction and various other miscellaneous uses.

There are three recognized species of *Sweitenia* having separate geographic distributions, viz., *S. mahogani* jacquin, *S. macrophylla* King and *S. humilis* Succ (10). Although these natural populations exist mainly in Central America, and parts of Asia and Africa, unfortunately, population pressures on the land, colonization programs, commercial logging and other deforestation programs have been depleting the natural forests of mahogany there and in other parts of the world. Besides, political uncertainties in these areas and demand for the wood in other parts of the world warrant examining other avenues for establishing and growing large numbers of these populations. Somatic cell genetics using tissue, cell and protoplast culture technology and methods of micropropagation and cloning offer an alternate route for mass propagation and maintenance of large populations. Investigations carried out on *in vitro* establishment of callus tissue and micropropagation on two species of mahogany, viz., *S. mahogani* Jacquin and *S. macrophylla* King are described.

M. R. Ahuja (ed.), Somatic Cell Genetics of Woody Plants, 147–153.

3. MATERIALS AND METHODS

3.1. Seed material

Seeds of S. mahogani were obtained from Sri Lanka and of S. macrophylla from Western Samoa. Seeds were germinated and planted in the greenhouse and they develop into vigorously-growing plants. The young plants were maintained in the greenhouse (Fig. 1).

3.2. Methods

3.2.1. Callus Initiation. Seeds were surface-sterilized in 10% Chlorox for 10 min and after several rinses, they were soaked in sterile distilled water overnight, since the seed coat is hard. The seed coats were removed the next day and they were surface-sterilized again in 10% Chlorox for 5 min. and washed several times before they were transferred to the agar culture medium. Aseptic 1 cm stem segments and squares of leaves (1 cm.) were also used to initiate callus.

3.2.2. Culture Media. A modified Murashige & Skoog (7) medium referred as MS medium was used. In addition to the normal salts, 30 mg/l of glutamine was added. 30 g/l of sucrose and Difco-Bacto agar (1% w/v) was used. for some experiments the woody plant medium (WPM) (6) was used. The culture medium was supplemented with 1.0, 2.0 or 5.0 mg/l of 2,4-dichlorophenoxyacetic acid (2,4-D) and 1 mg/l of -naphthalene acetic acid (NAA). In certain experiments, a combination of 1.0 mg/l of kinetin and 1.0 or 5.0 mg/l of 2,4-D were used. The pH of the medium was adjusted to 5.6 or 5.8 before autoclaving.

Cultures were maintained in a growth chamber at 25 $\pm$ 1°C with 12 hr/day illumination.

3.2.3. Micropropagation. The method based on growth of axillary buds from branch segments of plants (2) was employed. 2-3 cm segments of shoots containing an axillary bud were surface-sterilized in 10% Chlorox for 10 min and rinsed several times in sterile distilled water. They were planted on MS or WPM agar medium containing 1.0 mg/l of kinetin (K) or 4.0, 10.0 or 20.0 mg/l of 6-benzylaminopurine (BAP). In some experiments activated charcoal (0.2 or 0.5%, w/v) was added to the medium.

4. RESULTS

4.1 General growth response. The two species of mahogany, viz., S. mahogani and S. macrophylla essentially showed similar response to the culture conditions, viz., media, growth substances, etc. Also, the plant materials did not show any specific difference in response to the two culture medium, viz., MS or WPM. Therefore, most of the experiments were carried out with MS medium only.

4.2 Response of cotyledons. Explants of seeds without seed coat, viz., cotyledons with embryo when transferred to the culture medium produced callus within 2-3 weeks. The cotyledons exposed directly to the agar surface produced large proliferations of callus within 6 weeks (Fig. 2). These proliferations appeared soft and very white in color and further subcultures resulted in a loose "snow-flake" appearance. Microscopic observations of these cells revealed cells with very active protoplasm and contents. The cells appeared elongated ranging from 15-100 µ in length. The tissues were very loose; but attempts to initiate liquid cultures from these were not successful.

Callus initiation and further growth occurred in media containing combinations of 2,4-D (1.0, 2.0 or 5.0 mg/l) + kinetin (1.0 mg/l) or combinations of 2,4-D (2.0 mg/l) + NAA (1.0 mg/l). Subsequent routine subcultures of the callus has been carried out only on a culture medium containing 2,4-D (2.0 mg/l) + NAA (1.0 mg/l).

4.3. Response of stem segments and leaf tissue. Stem segments and pieces of leaf tissue responded within 1 week after transfer to the culture medium. Callus was evident from the cut end of stem segments and continued growth resulted in extensive callus proliferation (Fig. 3). Similarly one week after inoculation, callus originated from the cut margins of leaves exposed to the culture medium or often from leaf surfaces (Figs. 4 & 5). Further growth produced a soft, almost transparent, colorless callus or in some cases green-pigmented tissue, particularly those produced from the leaf surfaces. The stem segments and leaf pieces showed similar response to the various growth substance combinations as the cotyledons.

Routine subcultures of the callus tissue obtained from the cotyledons, stem segments and leaf pieces have been carried out every 6-8 weeks for the last 10 months. At present, in vitro organogenesis from callus tissue are pursued using various combinations and concentrations of growth hormones.

4.4 Progress in micropropagation. This method by which an axillary bud on the node of stem is stimulated to grow by the influence of specific growth hormone(s) and eventually made to root to form a plantlet, viz., micropropagation is now widely used in horticultural and forest-tree research particularly for mass propagation. Efforts in this laboratory have been fairly successful. Aseptic explants of 2-3 cm stem segments having an axillary bud were transferred to agar culture medium, maintaining an upright position in test tubes. Three different hormone combinations, viz., kinetin (1.0 mg/l) + NAA (0.1 mg/l), BAP (4.0 mg/l) and BAP (20.0 mg/l + 0.2% charcoal) were tried. Ten replicates for each of the treatments were made.

Figure 6 shows the effect of the three growth hormone treatments on the number of shoots which developed from buds after six weeks in culture. All the ten replicates grown in medium containing BAP (4.0 mg/l) or BAP (20.0 mg/l + 0.2% charcoal) produced shoots, i.e. 100% shoot growth whereas the response on medium containing kinetin (1.0 mg/l) + NAA (0.1 mg/l) was only 50%. Therefore, BAP appears to stimulate the best response.

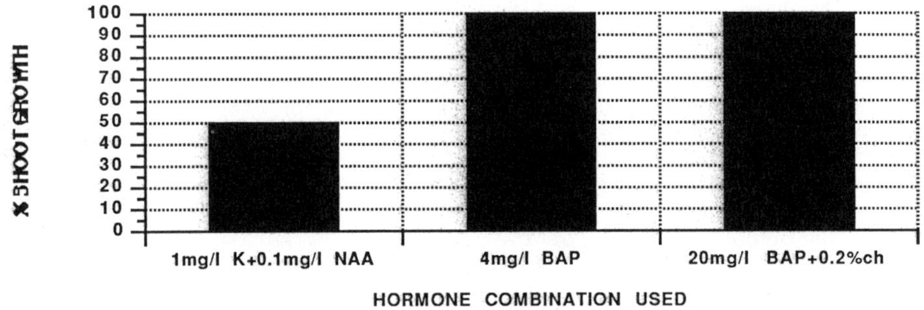

FIGURE 6. Number of shoots which developed from axillary buds of stem segments after six weeks in culture in various hormone combinations. Ten replicates were done for each hormone combination.

Response on the growth (elongation) of the shoot indicated that the average shoot length was maximum in the medium with BAP (20.0 mg/l + 0.2% charcoal), less in BAP (4.0 mg/l) and very little in medium containing kinetin (1.0 mg/l) + NAA (0.1 mg/l). Figure 7 illustrates these results and suggest that a high BAP concentration (20.0 mg/l) in the presence of charcoal (0.2%) evokes the best response. Other workers have shown similar effect of BAP (9).

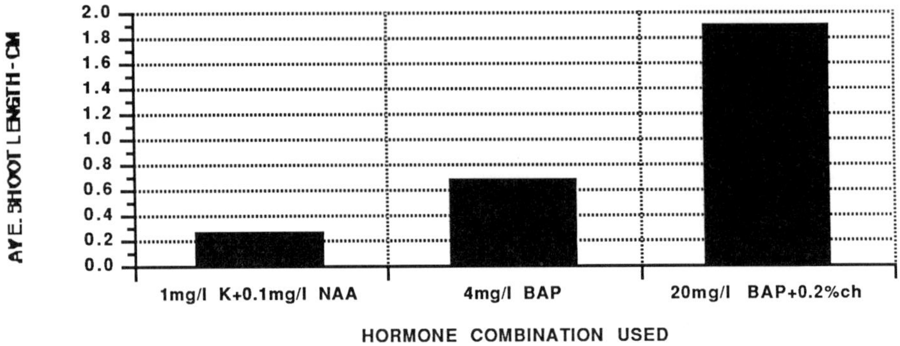

FIGURE 7. Average length of shoot growing out of the axillary bud of stem segments after six weeks in culture in various hormone combinations. Each bar represents average of ten replicates.

Figure 8 illustrates a stem explant with an axillary bud at the time of inoculation and Figs. 9-11 show the growth of the axillary bud into shoot in various hormone treatments.

Experiments are continued to root the elongated shoot to eventually form plantlets.

5. CONCLUSIONS AND PROSPECTS

These studies demonstrate that aseptic cell and tissue culture of the mahogany tree is feasible and successful. Segments of plant parts, viz., cotyledons, stem segments, leaves, etc. can be induced to form callus tissue and actively-proliferating tissues can be established and maintained for prolonged periods of time. The use and application of this tissue culture system of mahogany opens new avenues to study somatic cell genetics of this genus, viz., through regeneration from single cells, in vitro organogenesis leading to development of plantlets and eventually to selection of plants (trees) with improved qualities or resistance (3, 8).

In addition, the progress of research and results on micropropagation can be viewed optimistically as a practical and viable method for mass propagation of mahogany, particularly because of the decline in its natural populations. An added and unique advantage of this type of micropropagation is that the plant material can be kept in cold storage for a number of years employing the modern cryobiological preservation techniques. Other workers have examined this aspect in great detail for several gymnosperm and angiosperm genera, particularly for trees (1, 4, 5).

ACKNOWLEDGEMENTS

This research was partially supported by a NASA Grant, NAG 9-214 and in part from the University of Houston Coastal Center. We thank Dr. Glenn D. Aumann and Dr. Joe R. Cowles for their encouragement and support. The authors are grateful to Dr. S. Vivekanandan, Forest Department, Sri Lanka for supply of seeds of S. mahogany and the Department of Agriculture, Forests & Fisheries, Apia, Western Samoa for seeds of S. macrophylla. We thank Mrs. Kalyani Dias, Claudette Zaremba and Wanda Blackman for their technical assistance. An abstract on this work was presented at the poster session of the Symposium on Genetic Manipulations of Woody Plants, East Lansing, MI, USA, June, 1987.

6. REFERENCES

1. BAJAJ YPS, J REINERT 1977 Cryobiology of plant cell cultures and establishment of gene-banks. In J Rinert YPS Bajaj, eds, Applied and Fundamental Aspects of Plant Cell, Tissue and Organ Culture. Springer-Verlag Berlin, pp 757-777
2. BOULAY M 1987 Conifer micropropagation: Applied research and commercial aspects. In JM Bonga, DJ Durzan, eds, Cell and Tissue Culture in Forestry, Vol III, pp. 185-206
3. BONGA JM, DJ DURZAN 1982 Tissue Culture in Forestry. Martinus Nijhoff/Dr W Junk Publ., The Hague, 420 pp
4. HORGAN K 1987 Pinus radiata. In JM Bonga, DJ Durzan, eds, Cell and Tissue Culture in Forestry, Vol III, pp. 128-145
5. KARTHA KK 1986 Advances in cryopreservation technology of plant cells and organs. In CE Green, DA Somers, WP Hackett, DD Biesboer, eds, Plant Tissue and Cell Culture, pp. 447-458
6. LLOYD G, B McCOWN 1980 Commercially-feasible micropropagation of Mountain Laurel, Kalmia latifolia by use of shoot-tip culture. Proc Int Plant Propagators Soc 30: 421-427
7. MURASHIGE T, F SKOOG 1962 A revised medium for rapid growth and bioassays with tobacco tissue cultures. Physiol. Plant 15: 473-497
8. KARNOSKY DP 1981 Potential of forest tree improvement via tissue culture. Bioscience 31: 114-118
9. KIRKORIAN AD, RP KANN 1987 Aseptic multiplication of Chinese Tallow Tissue Culture (Sapium sebiferum (L.) Roxb.). In JM Bonga, DJ Durzan, eds, Cell and Tissue Culture in Forestry, Vol III, pp. 357-369
10. LAMB FB 1966 Mahogany of Tropical America, Univ. Michigan, Ann Arbor, Michigan, USA, 220 pp

FIGURE 1. *S. mahogani* plant growing in the greenhouse.
FIGURE 2. Proliferations of callus tissue from cotyledons of *S. mahogani* exposed directly to the agar surface of culture medium (MS) containing 2,4-D (2.0 mg/l) + NAA (1.0 mg/l) after six weeks in culture.
FIGURE 3. Callus tissue arising from stem explants of *S. mahogani* in MS medium containing 2,4-D (1.0 mg/l) + kinetin (1.0 mg/l) after 3 weeks in culture.
FIGURE 4. Callus from leaf segments of *S. mahogani* transferred to MS medium containing 2,4-D (1.0 mg/l) + BAP (1.0 mg/l). Note callus from cut regions and leaf surfaces.
FIGURE 5. Same as above of *S. macrophylla*, in MS medium containing 2,4-D (1.0 mg/l) + NAA (1.0 mg/l).

FIGURE 8. Stem explant with axillary bud inserted upright on agar medium (WPM) at the time of inoculation.
FIGURE 9. Response of axillary bud from stem explant after four weeks in medium (WPM) culture containing kinetin (1.0 mg/l) + NAA (0.1 mg/l).
FIGURE 10. Same as above in BAP (4.0 mg/l).
FIGURE 11. Same as above in BAP (20.0 mg/l) + charcoal (0.2%).

ATTEMPTS TO MICROPROPAGATE MATURE *LARIX DECIDUA* MILL.

J.M. BONGA and P. von ADERKAS, Canadian Forestry Service - Maritimes, P.O. Box 4000, Fredericton, N.B. Canada E3B 5P7.

ABSTRACT

Two, about 30-year-old *Larix decidua* trees were used in the experiments described. Female cones collected at about the time of meiosis and vegetative and sexual buds collected from late summer till early spring served as explants. Adventitious shoots developed either directly on these explants or on subcultured callus obtained from the explants. To induce the shoots the explants were exposed for 2 - 3 weeks to BA. Auxin in the medium was detrimental to all stages of development. For all types of explants the collection date was a critical factor in shoot initiation. Most shoots formed well developed needles and buds. However, only a small number of the adventitious shoots rooted and only a few of these showed stem elongation. Only one of the rooted shoots developed into a proper plant capable of continued growth after transfer to soil.

1. INTRODUCTION

Clonal propagation is becoming an important tool in reforestation (18, 30). To obtain the best clonal propagules for reforestation it is necessary that 1) the clonal parents are selected from sites with soil and climate conditions similar to those of the planting site, and 2) the clonal parents are old enough to have demonstrated that they are well adapted to the site, have the desired growth characteristics, and are disease resistant. At present, for proper selection of clones, it is generally required that the clonal parents are selected from trees of an age that is at least half that of trees at the end of their rotation, i.e., among mature trees (22).

Abbreviations: AP = allopurinol, BA = N^6-benzyladenine, CH = casein hydrolysate acid hydrolysed, 2,4-D = 2,4-dichlorophenoxyacetic acid, IBA = indolebutyric acid, 2iP = N^6-(2-isopentenyl)adenine, LM = Litvay et al. (19) medium, MCM = Bornman's (10) medium, NAA = naphthaleneacetic acid, QLP = 1/2 LM but with its microelements replaced by those of Quoirin and LePoivre's (24) medium, SH = Schenk and Hildebrandt's (25) medium, TIBA = triiodobenzoic acid, WPM = woody plant medium (20).

M. R. Ahuja (ed.), Somatic Cell Genetics of Woody Plants, 155–168.

Mature trees of most conifer species are difficult to clone by rooting of cuttings. Furthermore, even if rooting is obtained, the propagules often display plagiotropism and other temporary or persistent abnormalities (9, 29). Cloning of trees more than 10 years old by *in vitro* techniques often does not work well either, and has been reported only for *Pinus radiata* (17), and *Sequoia sempervirens* (3, 11). Explants from mature trees of *Cunninghamia lanceolata* (5) produced a large number of propagules but all of these were plagiotropic. Explants from mature trees of *Abies balsamea* (6), *Larix decidua* Mill. (7, 8), *Picea abies* (2), *P. glauca* (12), *P. pungens* (21), and *P. sitchensis* (26), *Pinus brutia* (1), *P. lambertiana* (13), and *Pseudotsuga menziesii* (28) formed adventitious shoots, but no or only a few, poorly growing rooted plantlets.

For several years we have been working with explants from mature *Larix decidua* Mill. and have reported the development of adventitious shoots from these explants. The adventitious shoots were obtained from the somatic tissues of female cones collected around the time of meiosis (8) and from vegetative and sexual buds collected in late summer (7). Over the last three years we have gathered additional information which is presented below.

2. MATERIAL AND METHODS

2.1. Collection and storage

All material was obtained from two, about 30-year-old *Larix decidua* trees. Because the explants of both trees behaved similarly *in vitro*, material from both trees was mixed randomly before explant excision. Two trees were used because enough material could not always be collected from one tree.

2.1.1. *Female cones at around the time of meiosis.* Meiosis, which lasts a few days, generally starts within a few days of the first of May. Therefore, collection of female cones was started each year in late April and was continued twice a week until mid May.

2.1.2. *Vegetative and sexual buds.* There are four types of buds on *Larix*; those that develop into long-shoots, short-shoots, female-cones, and male-cones. Twigs with these four types of buds were collected at irregular intervals between mid August and early April. Most of the twigs were used immediately after collection, but some were cold stored for later use. For cold storage 15- to 20-cm-long twigs were placed in 1 litre beakers with 12 g of activated charcoal and 200 ml of water. The charcoal was added to the water to absorb toxic secretions from the twigs. The beakers were subsequently covered with thin, clear plastic film with three, about 10-mm ventholes cut in it. These beakers with twigs were placed under 8/24 h of cool-white fluorescent light (50 μmol m^{-2} sec^{-1}) at 4C. Material thus stored was used after about 6 weeks of storage. Storage longer than 6 weeks resulted in reduced explant vigor and increased contamination rates.

2.2. Surface sterilization and excision

2.2.1. *Female cones at about the time of meiosis.* The female cones collected in April-May were surface sterilized in commercial bleach as described earlier (8). Unless otherwise stated, three transverse discs, about 1.5 mm thick, were cut from the sexual part of the cone, starting immediately above the stalk. The sexual part of the cone above the third slice was discarded. All excisions were made with the tissues submerged in malonic acid (8). In one experiment, the stalk as well as different components (Table 3) of the sexual part the female cone were excised and cultured.

2.2.2. *Vegetative and sexual buds.* Twigs with buds were cut in 2- to 3-cm long sections and surface sterilized by the same method as was used for the female cones. The bud scales were removed, and the primordial long-, short-, female-, and male-shoots were excised by a cut below the crown. All excisions were made with the tissues submerged in malonic acid (8).

2.3. Media and culture conditions

Unless otherwise indicated the medium used was that designed by Litvay et al. (19) but with all minerals, except FeEDTA, reduced to half (1/2 LM). For the first three weeks the medium contained 2 mg/l BA, but no auxin. Subsequently, the cultures were subcultured every 3-4 weeks on medium free of BA and auxin. The agar (Difco-Bacto) concentration in all media was 0.8%. The pH of the media was adjusted to 5.6 before autoclaving at 120C for 15 min. Fifteen millilitres of medium was used for each test tube (18 x 150 mm). The tubes were closed with "Morton" stainless steel caps and were placed upright in racks (40 tubes per rack). After receiving the explants each rack was enclosed in a clear plastic bag that was punctured in a few places to improve air circulation. The cultures were kept at a constant 21C under GRO-LUX (Sylvania F40T12-GRO) fluorescent lights 16/24 h at 50 mol m^{-2} s^{-1}.

3. RESULTS

3.1. Description of shoot and root development

3.1.1. *Female cones at about the time of meiosis.* Adventitious shoots developed either directly on the cut surface of the disc explants or from initial or subcultured callus of material collected over a short period in late April - early May (Tables 1-3), as described previously (8). In spite of many attempts to obtain stem elongation, stem length never exceeded 15 mm. Several methods commonly used for stem elongation, such as reducing the strength of the nutrient medium, adding charcoal to the medium, adding gibberellic acid to the medium, or exposing the shoots to far red light (29), were ineffective. Similarly, attempts to obtain rooting, either spontaneously or by treatment with an indolebutyric acid containing rooting powder (15), were not successful.

TABLE 1. Numbers of female cone explants that produced adventitious shoots on various nutrient media. Media contained BA 2 mg/l for the first three weeks of culture and no growth regulator thereafter. Forty explants per collection date per treatment

Nutrient medium	Collection date	
	May 3	May 7
1/2 LM	12	14
1/2 LM with QLP microelements	6	3
SH	8	9
WPM	6	3

TABLE 2. Numbers of female cone explants forming adventitious shoots after exposure to various growth regulators at various concentrations for various lengths of time. Forty explants per date per treatment. All explants were transferred to no-hormone medium after treatment

Treatment		Collection date			
		April		May	
Hormones (mg/l)	Duration (weeks)	26	30	3	7
BA 2	2	6	4	17	10
	4	2	0	9	4
	6	3	1	4	5
BA 2 + 2,4-D 0.01	4		0	0	
0.1	4		0	0	
+ NAA 0.01	4		0	0	
0.1	4		0	2	
+ TIBA 0.05	4		7	18	
BA 1 + 2iP 1	4		5	11	7
+ AP 1	4		8	13	7
10	4		8	22	13

TABLE 3. Numbers of explants that produced adventitious shoots. Explants are various sections of female cones collected around the time of meiosis. Forty explants per collection date per explant type

	Collection date					
	April	May				
Explant type	30	3	7	10	14	17
Slices of sexual part of cone	12	5	8	6	7	2
Slices of stem of cone	0	0	0	0	0	0
Bracts + o.s. + ovules	5	2	4	5	2	1
Bracts	7	2	0	1	0	1
O.s. + ovules	9	2	4	4	3	0
O.s.	0	0	0	3	0	2
Ovules	0	0	0	0	0	0

O.s. = Ovuliferous scale

3.1.2. Vegetative and sexual buds. Till late September it was difficult to determine if, at the time of excision, the explants were long-, short-, female-, or male- primordial shoots because of their small size. Therefore, in the initial experiments (Tables 4-6) no effort was made to identify them, i.e., random mixtures of the four shoot types were cultured. In later experiments (Table 7) the shoots were identified when excised.

In culture, most explants elongated a few millimetres and then often formed a green callus that soon covered most or all of the original explant. This callus developed mostly from the base of the elongating needles of the vegetative shoots, and bracts, and microsporophylls of the sexual shoots. In the long-shoot explants callus sometimes originated inside the apical dome and eventually burst through the outside cell layers of the dome. Some of the calli were granular, but most had a smooth surface. After about one month in culture, adventitious shoots started to develop from the smooth green callus in many of the cultures (Fig. 1).

Not all adventitious shoots were of callus origin. In the long-shoots some adventitious shoots developed from the surface of the expanding apical dome (Fig. 2). Others originated from needle primordia of the short- and long-shoots, and bract primordia of the female-shoots (Fig. 3). This development was similar to that described by von Arnold and Eriksson (2) for adventitious shoot development from needle primordia of *Picea abies*. A few adventitious shoots were noted on the abaxial surface of the elongating bracts of the female-shoots.

Within about two weeks of the first sign of adventitious shoot formation, dense clumps of shoots developed (Fig. 1). These shoots were separated as much as possible into single shoots, or small bunches of shoots and transferred to fresh medium. The callus maintained a steady rate of growth and kept producing new adventitious shoots, generally through three to four, but in a few cases up to nine or ten subcultures. Other shoots were obtained through lateral branching of the adventitious shoots.

Like the stems of the adventitious shoots produced by the female cones collected at about the time of meiosis, most of the stems of the adventitious shoots produced by the vegetative and sexual primordial shoots did not elongate more than a few millimetres. A few stems elongated to about 15 mm and two to about 40 mm (Fig. 4). One shoot treated with auxin containing rooting powder (15) and nine of the shoots that were not treated with auxin developed roots (Figs. 5-6). Seven of the rooted shoots were obtained from long-shoot explants, three from female-shoot explants; eight were from September and two from March collections. All except one (Fig. 6) of the shoots had stems only a few millimetres long when they rooted. All rooted shoots were transferred to soil. Nine of the rooted shoots lived for several months without displaying further stem elongation and then died. All had a well developed apical bud and lateral buds but these remained dormant, even after a 6-week cold treatment (4C and 8/24 h 50 μmol m^{-2} s^{-1} cool-white fluorescent light). In the tenth rooted shoot the stem elongated rapidly and a plant (Fig. 7) similar in appearance to a normal seedling developed. At the stage shown in Figure 7 the plant was transferred to a bigger pot. During the transfer, it was noted that the plant had two main roots about 15 cm long. Shortly after transfer the plant went dormant. After a 6-week cold treatment it flushed and is now developing lateral shoots, although these appear to be only short-shoots.

3.2. The effect of some experimental variables

3.2.1. Recording of data. Often it was difficult or impossible to determine the exact number of adventitious shoots per explant or callus, because 1) the shoots frequently appeared in dense masses that were difficult to disentangle, 2) it was often difficult to distinguish between shoots with a proper apical bud and mere bunches of adventitious needles. Therefore, our data are based on the number of explants or their callus that produced adventitious shoots, rather than on the total number of shoots produced. The data were tested statistically by analysis of variance (27).

3.2.2. Experiments with female cones collected when at or near meiosis. In an earlier publication (8) we reported on the difference in response of cultures on different basal media. Of the media tested, 1/2 LM was the best for adventitious shoot induction. Since then we have tested QLP, SH, and WPM. Of these, our standard 1/2 LM was the most satisfactory (Table 1). The results obtained with this medium were significantly different (p = $<$0.001) from those obtained with the other three media.

In our earlier experiments (8) it was also found that addition of IBA to the medium inhibited adventitious shoot formation. In a later experiment (Table 2) we tested the effects of 2,4-D because it often induces embryogenesis (14), and NAA because it is a relatively stable auxin that is generally milder in its action than IBA. We also tested the antiauxin TIBA, exposure to BA for different lengths of time, and exposure to combinations of BA and the cytokinins 2iP and AP. Addition of 2,4-D and NAA strongly inhibited shoot induction, and the response to TIBA was similar to that of a 2-week exposure to BA. Two week exposure to BA was not significantly more effective than 4- or 6-week exposure. The response on the BA plus 2iP combination was similar to that of the short-term exposure to BA. A high response was obtained with BA plus AP, but the shoots produced by this combination were highly stunted. The material collected May 3 was significantly more responsive (p = $<$0.01) than the April 30 material.

In a third experiment, the sexual cones were cut into their various component parts (Table 3). The transverse discs cut from the sexual part of the cone, i.e., the type of explant used routinely in our experiments, were significantly more responsive ($p = < 0.001$) than the other explant types. Sections of the stalk of the cone did not respond at all, and response was also poor in ovuliferous scale (minus ovule) and ovule explants.

3.2.3. Experiments with vegetative and sexual buds. In the first experiment (1982-1983) material from three collection dates during the predormant period and from one collection in January were used (Table 4). Only the two collections in September responded.

The next year (1983-1984), it was decided to investigate when during the predormant period the material is most active. Therefore, material was collected weekly from August 11 till September 26. Again, a January collection was also made. In addition, MCM medium, a medium previously tested by us with female cones collected at about the time of meiosis (8), was compared with our usual 1/2 LM medium. Furthermore, the experiment was repeated with material that had been in cold storage for six weeks (Table 5). There was little response in material collected before the end of August; the response was strong in material collected throughout September, and nil in the January material. Significantly ($p = < 0.025$) more explants formed adventitious shoots on 1/2 LM than on MCM medium.

In 1985-1986 the emphasis was on the winter season (Table 6), with only one collection date (September 9) in the predormant period. The response was best in September and the end of March.

In 1986-1987, the differences in response between long-, short-, female-, and male-shoot explants was determined in material collected during the two active periods, i.e., September and March (Table 7). The response was significantly higher ($p = < 0.01$) in the long-shoot explants than in the others; it was poorest in the male-shoot explants. The general response was better in the end of March material than in the early March material.

In another experiment, the effect of casein hydrolysate and glutamine supplemented in various proportions was studied (Table 8). The supplements did not stimulate shoot induction over the shoot induction level obtained in the control treatment.

4. DISCUSSION

The results obtained with female cones collected at about the time of meiosis confirmed earlier observations (8) that 1/2 LM is a good medium for *in vitro* culture of mature *Larix decidua*, and that these cultures are very sensitive to auxin. In addition, it was shown that BA is required for only a short period (2 weeks) for adventitious shoot induction. We routinely use a 3-week exposure to BA, which generally results in healthy, non-vitrified shoots with long (20-40 mm) needles. In our routine procedure we use discs from the sexual part of the cone. These explants

TABLE 4. Numbers of shoot explants that produced adventitious shoots. Two hundred and forty explants per collection date

August	September		January
23	3	18	10
0	9	51	0

TABLE 5. Numbers of shoot explants that produced adventitious shoots. Effect of collection date, storage, and nutrient media. With fresh material eighty explants per collection date per nutrient medium; with cold stored material forty explants per collection date per nutrient medium

Cold storage	Medium	August				September				January
		11	15	22	29	6	12	19	26	6
No	1/2 LM	0	0	15	9	26	13	6	16	0
No	MCM	0	0	4	0	8	0	0	0	0
Yes	1/2 LM	3	2	3	4	0	1	0	6	0
Yes	MCM	1	0	0	0	0	0	0	3	0

TABLE 6. Numbers of explants that produced adventitious shoots. Effect of collection date. Forty explants per collection date

Collection date						
September	November	February			March	
9	22	11	18	26	18	28
8	1	2	0	1	2	7

TABLE 7. Numbers of shoot explants that produced adventitious shoots. Effect of collection date and shoot type. Eighty explants per date per shoot type

Collection date		Shoot type			
		long	short	female	male
Sept	15	14	5	6	0
March	3	9	4	1	0
	24	12	0	0	2
	30	29	8	4	0

TABLE 8. Numbers of explants that produced adventitious shoots on media supplemented with various concentrations (mg/l) of casein hydrolysate and glutamine. Forty explants per collection date per treatment

Supplement		Collection date		
CH	Glutamine	Aug 22	Sept 5	Sept 19
0	0	0	3	1
100	0	1	1	1
300	0	1	1	3
900	0	2	0	0
0	100	0	0	2
0	400	2	0	0
900	400	3	4	1

may be more responsive than excised portions thereof because of their larger size. With a larger volume/cut surface ratio the wounding damage presumably is less, which could explain the better growth response. In these disc explants, the vast majority of the cells are somatic diploid cells, the haploid tissue being confined to one tetrad, or at most one small coenocyte per ovule. Development of tissues from these haploid cells was never noticed; all adventitious shoots were of diploid origin.

The experiments with vegetative and sexual primordial shoots showed that the explants were morphogenetically more active before the onset of the dormant season and just before budbreak than during the fall and winter. Increased morphogenetic activity has also been observed in cultures of buds of *Abies balsamea* (6), and *Picea sitchensis* (26) collected shortly before budbreak. During late summer and in the spring, various kinds of primordia are initiated in the vegetative and sexual buds (16, 23). It appears therefore, that in tissues collected during these short periods of increased *in situ* morphogenetic activity, the morphogenetic plasticity *in vitro* is improved as well (9).

The cultures are sensitive to auxin, exogenous auxin being inhibitory during the adventitious shoot induction phase as well as during the rooting phase. At present, we do not know if this sensitivity is a characteristic of cultures of explants from mature trees, or if this sensitivity is associated with the species.

In our cultures we have obtained a large number of adventitious shoots. These shoots have healthy, normal looking needles and buds. However, almost all of these buds remain dormant, even after a 6-week exposure to short day light at 4C. Possibly it is this bud dormancy that also prevents stem elongation in these shoots. It has been noted by Bekkaoui et al. (4) that when shoot producing cultures of *Pseudotsuga menziesii* start to degenerate after several subcultures, the apical buds in the shoots become dormant. They also observed that these buds had accumulated large amounts of zeatin and zeatin-riboside. At present it is not known if the lack of bud flushing and stem elongation in our cultures is similarly caused by excessive cytokinin accumulation in the buds.

Most of the rooted shoots also failed to flush and elongate, even though these shoots are probably well supplied with nutrients taken up by the roots. It is possible that nutrient transport from the roots was impeded by a lack of proper vascular connections between the roots and the shoots. However, the fact that all the roots observed formed in areas of the shoots with little or no callus, and that the rooted shoots remained fresh-green for six to seven months after transfer to soil, indicate that vascular connections were well established.

With only one rooted shoot elongating properly, we obviously still are a long way from a functional micropropagation system. Nevertheless, the procurement of this one plantlet is encouraging because it shows that micropropagation is potentially possible.

At present there is a strong desire to obtain micropropagation of selected superior mature individuals. However, this may not remain the case. If methods are developed whereby the potential for growth and other characteristics can be established at an increasingly younger age of the individual, the need for micropropagation of older individuals will diminish.

5. REFERENCES

1. Abdullah AA, Yeoman MM, Grace J (1987) Micropropagation of mature Calabrian pine (Pinus brutia Ten.) from fascicular buds. Tree Physiol 3: 123-136.
2. Arnold S von, Eriksson T (1979) Induction of adventitious buds on buds of Norway spruce (Picea abies) grown in vitro. Physiol Plant 45: 29-34.
3. Ball EA, Morris DM, Rydelius JA (1978) Cloning of Sequoia sempervirens from mature trees through tissue culture. In: Proc Round Table Conf "In Vitro" Multiplication of Woody Species. Gembloux (Belgium) June 6-8, Centre de Recherches Agron de l'Etat. p 181-226.
4. Bekkaoui F, Maldiney R, Pilate G, Boulay M, Franclet A (1986) Dégénerescence clonale du douglas (Pseudotsuga menziesii) cultivé in vitro; relation avec la teneur en éléments minéraux, en auxine, acide abscissique, zéatine et zéatine-riboside. C R Acad Sc Paris 303: 13-18.
5. Bigot C, Engelmann F (1987) Vegetative propagation in vitro of Cunninghamia lanceolata (Lamb.) Hook. In: Bonga JM, Durzan DJ (eds) Cell and Tissue Culture in Forestry, Volume 3, Case Histories: Gymnosperms, Angiosperms and Palms. Martinus Nijhoff/Dr W Junk Publ, Dordrecht p 114-127.
6. Bonga JM (1981) Organogenesis in vitro of tissues from mature conifers. In Vitro 17: 511-518.
7. Bonga JM (1984) Adventitious root and shoot formation in tissue cultures of mature Larix decidua. In: Hanover J, Karnosky D, Keatly D (eds) International Symposium of Recent Advances in Forest Biotechnology, Traverse City, Michigan p 64-68.
8. Bonga JM (1984) Adventitious shoot formation in cultures of immature female strobili of Larix decidua. Physiol Plant 62: 416-421.
9. Bonga JM (1987) Clonal propagation of mature trees: problems and possible solutions. In: Bonga JM, Durzan DJ (eds) Cell and Tissue Culture in Forestry, Volume 1, General Principles and Biotechnology. Martinus Nijhoff/Dr W Junk Publ, Dordrecht p 249-271.
10. Bornman CH (1983) Possibilities and constraints in the regeneration of trees from cotyledonary needles of Picea abies in vitro. Physiol Plant 57: 5-16.
11. Boulay M (1987) Conifer micropropagation: applied research and commercial aspects. In: Bonga JM, Durzan

DJ (eds) Cell and Tissue Culture in Forestry, Volume 3, Case Histories: Gymnosperms, Angiosperms and Palms. Martinus Nijhoff/Dr W Junk Publ, Dordrecht p 185-206.

12. Dunstan DI, Mohammed GH, Thorpe TA (1987) Morphogenetic response of vegetative bud explants of adolescent and mature *Picea glauca* (Moench) Voss *in vitro*. New Phytol 106: 225-236.
13. Gupta PK, Durzan DJ (1985) Shoot multiplication from mature trees of Douglas fir (*Pseudotsuga menziesii*) and sugar pine (*Pinus lambertiana*). Plant Cell Rep 4: 177-179.
14. Halperin W (1986) Attainment and retention of morphogenetic capacity *in vitro*. In: Vasil I (ed) Cell Culture and Somatic Cell Genetics of Plants, Volume 3, Plant Regeneration and Genetic Variability. Academic Press Inc, Orlando p 3-47.
15. Hare RC (1978) Effect of shoot girdling and season on rooting of slash pine cuttings. Can J For Res 8: 14-16.
16. Hashizume H, Imai M (1966) On the developmental process of flower buds in *Larix leptolepis*. J Japan For Soc 48: 425-435.
17. Horgan K (1987) *Pinus radiata*. In: Bonga JM, Durzan DJ (eds) Cell and Tissue Culture in Forestry, Volume 3, Case Histories: Gymnosperms, Angiosperms and Palms. Martinus Nijhoff/Dr W Junk Publ, Dordrecht p 128-145.
18. Libby WJ (1979) Clonal selection of forest trees. Calif Agric 33: 8-9.
19. Litvay JD, Johnson MA, Verma D, Einspahr D, Weyrauch K (1981) Conifer suspension culture medium development using analytical data from developing seeds. Inst Paper Chem Tech Paper Ser 115: 1-17.
20. LLoyd GB, McCown BH (1980) Commercially feasible micropropagation of mountain laurel (*Kalmia latifolia*) by use of shoot tip culture. Proc Intern Plant Propag Soc 30: 421-437.
21. Misson JP, Coumans M, Giot-Wirgot P, Gaspar TH (1982) Induction de bourgeons adventifs sur bourgeons de *Picea pungens* en culture *in vitro*. Z Pflanzenphysiol 107: 161-167.
22. Namkoong G, Barnes RD, Burley J (1980) Screening for yield in forest tree breeding. Commonw For Rev 59: 61-68.
23. Owens JN, Molder M (1979) Bud development in *Larix occidentalis*. I. Growth and development of vegetative long shoot and vegetative short shoot buds. Can J Bot 57: 687-700.
24. Quoirin M, LePoivre P (1977) Etude de milieux adaptes aux cultures *in vitro* de *Prunus*. Acta Hortic 78: 437-442.
25. Schenk RU, Hildebrandt AC (1972) Medium and techniques for induction and growth of monocotyledonous and dicotyledonous plant cell cultures. Can J Bot 50: 199-204.
26. Selby C, Harvey BMR (1985) The influence of natural and *in vitro* bud flushing on adventitious bud production in

Sitka spruce (Picea sitchensis (Bong.) Carr.) bud and needle cultures. New Phytol 100: 549-562.
27. Sokal RR, Rohlf FJ (1981) Biometry. WH Freeman and Company, San Fransisco, 2nd edition.
28. Thompson DG (1981) Induction of adventitious buds on cultured embryonic shoots of Douglas fir Pseudotsuga menziesii (Mirb.) Franco. Environ Exp Bot 21: 433.
29. Thorpe TA, Biondi S (1984) Conifers. In: Sharp WR, Evans DA, Ammirato PV, Yamada Y (eds) Handbook of Plant Cell Culture, Volume 2, Crop Species. Macmillan Publishing Company, New York p 435-470.
30. Zobel B (1981) Vegetative propagation in forest management operations. In: Proceedings 16th Southern Forest Tree Improvement Conference. Virginia Polytechnic Institute and State University, Blacksburg, VA p 149-159.

6. FIGURES

Figure 1. A primordial female-shoot explant, excised from a bud collected in September, has become completely overgrown with callus and adventitious vegetative shoots. Bar = 1 mm.

Figure 2. Adventitious primordia (arrows), some of which will develop into shoots and some into needles, have been initiated on the surface of a swollen apical dome of a primordial long-shoot explant. The explant was excised from a bud collected in September. Bar = 1 mm.

Figure 3. A bract primordium of a primordial female-shoot explant, excised from a bud collected in March, is turning into a vegetative adventitious shoot. Bar = 1 mm.

Figure 4. A section of a 40 mm long adventitious shoot that had arisen from subcultured callus from a primordial male-shoot explant, excised from a bud collected in March. Note lateral bud (arrow). Bar = 1 mm.

Figure 5. A rooted adventitious shoot that had arisen from a callusing primordial shoot (type not determined) explant, excised from a bud collected in September. The stem of this shoot did not elongate in spite of the presence of a well developed apical bud and root system. Bar = 10 mm.

Figure 6. A rooted adventitious shoot that had arisen from subcultured callus from a primordial shoot (type not determined) explant, excised from a bud collected in September. The stem of this shoot had elongated to 15 mm. The position of the apical bud, hidden behind some needles, is indicated by the arrow. Bar = 10 mm.

Figure 7. A vigorously growing plantlet obtained from a rooted adventitious shoot that had arisen from subcultured callus from a primordial female-shoot explant, excised from a bud collected in March. Bar = 10 mm.

1
2
3
4
5
6
7

PROTOPLAST CULTURE OF WOODY ROSACEAE AND A COMPARISON TO HERBACEOUS ROSACEAE

JÖRG JÖRGENSEN AND HORST BINDING

Niedersächsiche Forstliche Vesuchsanstalt, 3513 Staufenberg/ Escherode, und Botanisches Institut und Botanischer Garten, Christian-Albrechts Universität, 2300 Kiel, Federal Republic of Germany

0. ABSTRACT

Protoplasts from a number of woody and herbaceous species of Rosaceae were cultured to investigate their growth potential. In the woody species, protoplasts from shoot tips with small but unexpanded leaves could be successfully cultured; whereas in the herbaceous species, protoplasts from fully expanded leaves were also capable of cell divisions. Colony formation and callus development have been observed in the species of Rosaceae.

1. INTRODUCTION

Generally protoplasts derived from the species of the family Rosaceae are very difficult to grow. Only in a few species could the plants be regenerated from the protoplasts. These include: Fragaria ananassa (3), Prunus dulcis (12), and Pyrus communis (9). In Rosa (7, 11) and Malus (6) only callus could be regenerated from the protoplasts.

2. MATERIALS AND METHODS

In our investigations with four woody species, including a hybrid, and six herbaceous species of Rosaceae, we were able to regenerate callus from the protoplasts cultures.The woody species included: Sorbus aucuparia, Prunus avium , P. cerasus, and P. avium x P. cerasus "Colt"; the herbaceous species included: Agrimonia eupatoria, Duchesnea chrysantha, Potentilla alchemilloides, P. caulescens, P. erecta, and P. frigida.

The protoplasts were obtained from shoot tips harvested from shoot cultures maintained on MS (8) or B5 (4) media, supplemented with 2.5 µM BAP. The shoot tips were cut into slices in a solution containing 0.5 M mannitol (an osmoticum), 0.25 % PVP and 2.5 mM diethyldithiocarbamic acid , in order to control the phenolic compounds.

The slices were washed and incubated for 8 to 16 hours in digestion mixture containing 0.5 to 1.0 % macerozyme, 1.0 to 3.0 % meicillase or cellulysen, 5 mM ca(NO3)2, 0.5 to 1.0 g/l PVP, and 0.5 M mannitol, Protoplasts were centrifuged, washed and cultured in agarose droplets (1, 10) in V-KM medium (2) supplemented with 2.5 µM BAP, 5.0 µM NAA and 0.5 µM 2,4-D.

3. RESULTS AND DISCUSSION

After two to 6 days, first cell divisions (Figure 1) were observed. After 14 days, cell colonies (Figure 2) had reached diameters up to 1 mm. Callus cultures of Sorbus aucuparia (5), Agrimonia eupatoria, Duchesnea chrysantha, Potentilla alchemi-

M. R. Ahuja (ed.), Somatic Cell Genetics of Woody Plants, 169–172.

lloides, _P. caulescens_, _P. erecta_ and _P. frigida_ could be transfered to B5 medium with 2.5 µM BAP and 50 ml/l coconut water. The calluses of _Prunus_ species (Figures 3 and 4) only survived if transfered to V-KM media following stepwise reduction of osmoticum ranging from 0.38 to 0.10 M glucose. The callus growth continued on B5 medium with 2.5 µM BAP and 50 ml/l coconut water.

A comparison between woody and herbaceous species of Rosaceae shows that, as far as woody plants are concerned, only protoplasts from shoot tips with small, not expanded leaves could be successfully cultured; whereas from the herbaceous species fully expanded leaves were also suitable source of protoplast culture. It was easier to get high degrees of dividing protoplasts from herbaceous plants, but only in a few species callus was obtained. In all tested species of Rosaceae, the phenolic compounds are a great problem. Thus viable protoplasts could be reliably isolated only by cutting shoot tips and leaflets in solutions with PVP alone, or PVP and diethyldithiocarbamic acid, which protect the cells from phenolic compounds. Best results were obtained in all species tested by culture of protoplasts in agarose droplets, as compared to liquid cultures.

REFERENCES

1. Binding H: Mitteilungsband - Kurzfassungen der Beiträge der Botaniker-Tagung, Wien, p. 61, 1984.
2. Binding H and R Nehls: Z Pflanzenphysiol 85: 279, 1977.
3. Binding H et al: In: Proc 5th Int Congr Plant Tissue & Organ Culture, Tokyo, p. 575, 1982.
4. Gamborg OL et al: Exp Cell Res 50: 151, 1968.
5. Jörgensen J and H Binding: Z Pflanzenphysiol 113: 371, 1984.
6. Koudier M et al: Plant Cell Rep 3: 142, 1984.
7. Krishnamurthy KV et al: Plant Sci Lett 15: 135, 1979.
8. Murashige T and F Skoog: Physiol Plant 15: 473, 1962.
9. Ochatt SJ and OH Caso: J Plant Physiol 122: 243, 1986.
10. Shepard JF: Plant Sci Lett 18: 327, 1980.
11. Strauss A and I Potrykus: Physiol Plant 48: 15, 1980.
12. Wu S and A Kuniyuki: Plant Sci Lett 41: 55, 1985.

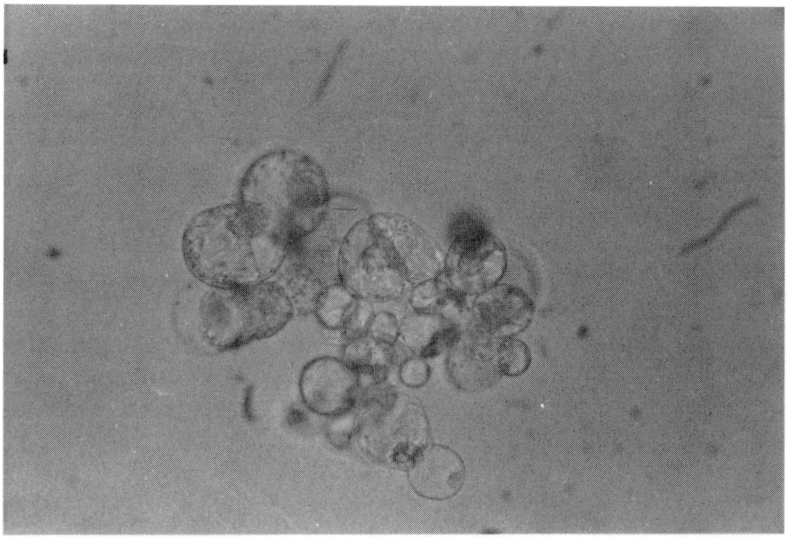

Figure 1. Cell division in protoplast culture of *Sorbus aucuparia*

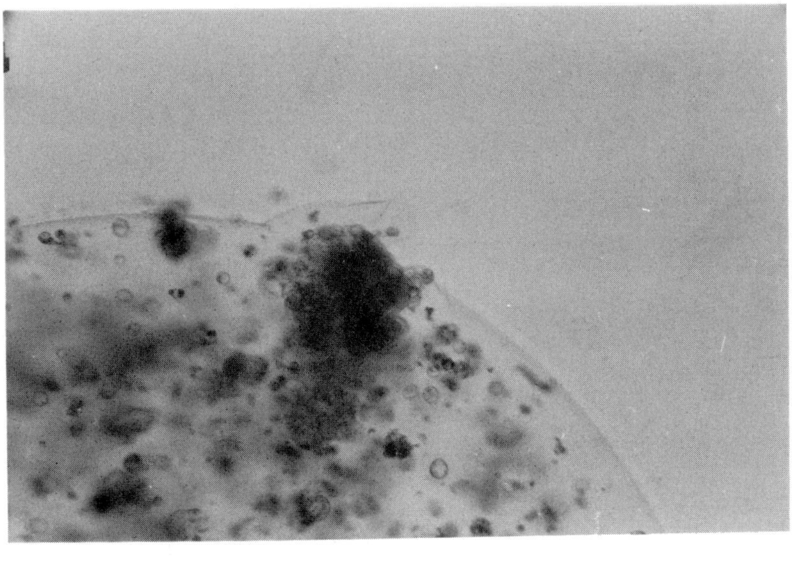

Figure 2. Cell clusters (*Prunus avium* x cerasus "Colt")

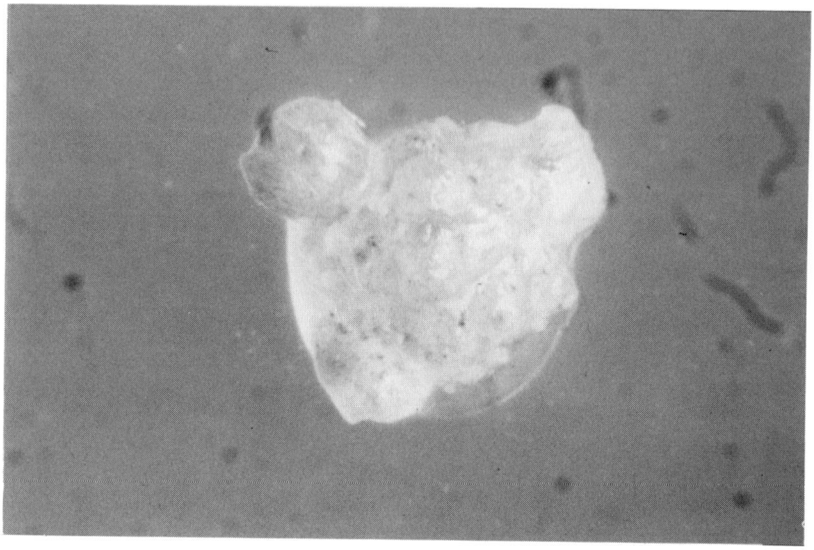

Figure 3. Small callus in agarose (*Prunus cerasus*)

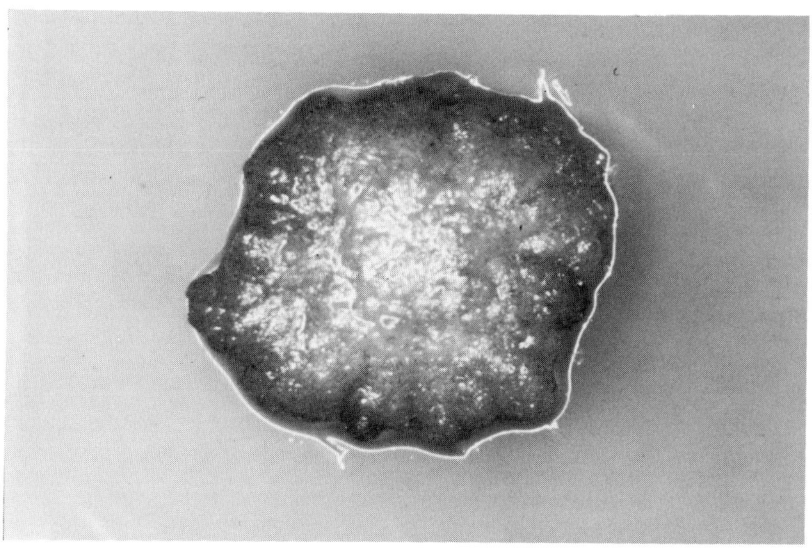

Figure 4. Callus on MS medium with coconut water (*Prunus avium*

DIFFERENTIAL GROWTH RESPONSE OF ASPEN CLONES STORED AT SUB-ZERO TEMPERATURES

M. R. AHUJA

Federal Research Centre for Forestry and Forest Products, Institute of Forest Genetics and Forest Tree Breeding, Sieker Landstrasse 2, D-2070 Grosshansdorf, Federal Republic of Germany

0. ABSTRACT

The potential of dormant buds for conservation of forest tree germplasm was investigated. Dormant buds along with twigs from 14 aspen (*Populus tremula*, *P. tremuloides*) and hybrid aspen clones were wrapped in plastic bags and stored at five different sub-zero temperatures (0 C, -5 C, -8 C, -18 C, and -80 C) for up to two years. In another set of experiments, only dormant buds from two aspen and two hybrid aspen clones were stored in screw-cap bottles or sealed ampules at the five sub-zero temperatures for up to two years. Before storage, some dormant buds were coated with wax and stored under vacuum. After different periods of storage, bud explants were cultured on a screening medium to monitor the growth and differentiation of microshoots. Results of these studies show that storage potential of an aspen clone is largely determined by the storage temperature, the genotype, and the conditions of storage. Of the five temperatures tested, -80 C seemed to be the best temperature for long-term storage of dormant buds.

1. INTRODUCTION

The state of forest tree gene resources is reaching a danger point due to cutting down of forests for agricultural and industrial use, energy, roadways, human settlements, and environmental pollution, for example "Waldsterben" (dying forests). In the Federal Republic of Germany more than 50 percent of the forests are dying or are already dead (5). Therefore, preservation of forest tree germplasm has become an important program area of forest tree genetics, breeding and improvement. Broadly speaking, the preservation of germplasm (7) may be carried out by:

- Preservation of forest tree gene resources *in situ*, which is very important in forest trees (for example, stands conserved by natural regeneration), and
- Preservation of germplasm *ex situ*, which offers some possibilities. These include evacuation of plantations, seed and clonal orchards, and storage of germplasm through seed, pollen, part of the plant, such as twigs, buds, meristems, tissues and cells.

Seed of the forest trees are generally stored at low temperatures ranging from 0 C to -20 C. In a recent study, the seed from a number of forest tree species were stored in liquid nitrogen (-196 C) and were tested for their viability after the storage period (3). The pollen grains of tree species have been

M. R. Ahuja (ed.), Somatic Cell Genetics of Woody Plants, 173–180.

stored for a number of years under low temperatures (-18 to -20 C), with or without vacuum (4, 6). Winter hardy twigs from several tree species, for example willow, poplar, and apple were stored in liquid nitrogen and were found alive after the storage (8).

The present study was undertaken to investigate the differentiation potential of dormant buds of aspen and hybrid aspen clones stored at sub-zero temperatures over a period of two years.

2. MATERIALS AND METHODS

2.1. Species and storage temperatures

Dormant buds collected in December-January, 1985 from European aspen (*Populus tremula*), quaking aspen (*P. tremuloides*), and hybrid aspen (*P. tremula* x *P. tremuloides*) clones were stored at five different sub-zero temperatures: 0 C, -5 C, -8 C, -18 C, and -80 C. Two different approaches were employed for storage of dormant buds.

2.1.1. Storage of twigs. In one set of experiments, twigs with dormant buds from 14 aspen and hybrid aspen clones were wrapped in plastic bags and stored at the five temperatures for up to two years.

2.1.2. Storage of buds. In another set of experiments, only dormant buds from two aspen (W 29, W 80) and two hybrid aspen (Esch 2, Esch 8) clones were stored at the five sub-zero temperatures under following conditions:

- Buds coated with wax and stored under vacuum (190 Torr) in sealed ampules.
- Buds coated with wax and stored without vacuum in screw-cap bottles.
- Buds without wax coating and stored under vacuum in sealed ampules.
- Buds without wax coating and stored without vacuum in screw-cap bottles.

2.1.3. *In vitro* culture. After different periods of storage, bud explants were cultured on a screening medium, modified Woody Plant Medium (1), to monitor the survival and growth and differentiation of microshoots. At least 8 to 10 bud explants were cultured per treatment. Results are expressed as mean number of microshoots per explant per clone, along with standard errors. The methods for the preparation of explants and cultural conditions were essentially the same as previously described (2).

3. RESULTS AND DISCUSSION

3.1. Storage of twigs

Of the 14 clones stored at five different temperatures for any length of time, bud explants from 12 clones were cultured before storage to determine the degree of microshoot differentiation on the primary explants (Tables 1, 2). Two *P. tremula* clones, W 3 and W 7, which produced more than 5 microshoots per explant (Table 1) were arbitrarily classified as 'high shooters', while the remaining 10 clones that produced less than 3 microshoots per explant were grouped as 'low shooters'

Table 1. Number of microshoots per bud explant in European aspen, *Populus tremula*, clones (W 3, W 7, W 25, W 97) and quaking aspen, *P. tremuloides*, clones (Turr 141, Schreiner) stored for one week at 0 C. (Based on 8-10 explants per clone).

W 3	W 7	W 25	W 97	Turr 141	Schreiner
5.1±0.8	5.4±0.8	0.9±0.3	1.7±0.4	0.4±0.2	1.3±0.4

Table 2. Number of microshoots per bud explant in hybrid aspen (*P. tremula* x *P. tremuloides*) clones for one at 0 C. (Based on 8-10 explants per clone).

Se 1	Se 9	Se 11	Esch 3	Esch 9	Esch 11
1.2±0.4	0.9±0.3	2.5±0.5	1.0±0.3	1.2±0.3	1.4±0.4

(Tables 1, 2).

3.1.1. Storage for one month. In order to determine the effect of five temperatures, bud explants from two clones, one a 'high shooter' (W 7) and the other a 'low shooter' (W 97), were cultured after one month of storage. Both clones showed differentiation of microshoots on all temperatures according to their potentialities (Table 3). Clone W 7 showed that the degree of microshoot differentiation was relatively higher at 0 C to -8 C temperatures as compared to -18 and -80 C (Table 3). Clone W 97, a 'low shooter' exhibited differentiation of microshoots at all five temperatures, within its norm.

Table 3. Degree of microshoot differentiation in explants taken from dormant buds stored with twigs for one month at five sub-zero temperatures (Based on 10 explants per treatment).

Clone	0 C	-5 C	-8 C	-18 C	-80 C
W 7	4.6±0.7	3.6±0.6	4.5±0.7	1.7±0.4	1.9±0.4
W 97	1.9±0.4	1.8±0.4	1.2±0.3	1.2±0.3	2.5±0.5

3.1.2. Storage for one year. The survival and capacity of the 14 clones to differentiate microshoots after storage for one year at five temperatures are shown in Table 4. The aspen and hybrid aspen clones responded quite differently to the five sub-zero temperatures. All six European aspen, *P. tremula*, clones showed extreme sensitivity to -18 C; the bud explants turned brown following culture, and did not grow or differentiate, except for very slight differentiation in W 7 and W 25 (Table 4). Only two clones, W 25 and W 97, showed some differentiation at 0 C, the remaining four clones (W 3, W 7, W 14, W 90) showed no growth and differentiation at 0 C. All six *P. tremula* clones showed differentiation of microshoots at -5 C, -8 C and -80 C, although there were significant differences between temperatures in three of 6 clones (Table 4). There seems to be an optimum for microshoot differentiation at -5 C.

Both the *P. tremuloides* clones (Turr 141 and Schreiner) showed practically no growth and differentiation at the five storage temperatures after one year. Only Schreiner showed very slight differentiation of microshoots at -80 C (Table 4).

Bud explants from the six hybrid aspen clones exhibited different responses to the five sub-zero temperatures after one year of storage. Each clone exhibited a differential response to the low temperatures tested (Table 4). Four of 6 clones (Se 1, Se 10, Esch 3, Esch 11) showed maximum number of microshoots at -80 C; the remaining two clones (Se 9 , Esch 9) exhibited practically no differences in the number of microshoots produced after storage at different temperatures on which they survived. Both these clones (Se 9, Esch 9) and

Esch 11 did not exhibit any growth at -18 C (Table 4). Two of 6 clones (Esch 3, Esch 9) also showed sensitivity to 0 C, and did not grow or differentiate microshoots at this temperature after one year storage (Table 4). There seeems to an optimum for microshoot differentiation at -80 C.

Table 4. Degree of microshoot differentiation in explants taken from dormant buds stored with twigs for one year at five sub-zero temperatures. (Based on 8-10 explants per treatment).

Clone	0 C	-5 C	-8 C	-18	-80 C
W 3	0	1.1±0.3	4.2±0.7	0	0.6±0.2
W 7	0	9.4±1.1	2.1±0.5	0.1	2.4±0.5
W 14	0	6.7±0.9	4.3±0.7	0	1.2±0.4
W 25	0.4±0.2	0.1	1.1±0.4	0.4	1.0±0.3
W 90	0	1.0±0.3	1.2±0.4	0	1.1±0.3
W 97	0.9±0.3	1.2±0.4	0.2	0	1.2±0.4
Turr 141	0	0	0	0	0
Schreiner	0	0	0	0	0.2
Se 1	0.1	1.4±0.4	1.2±0.4	1.4±0.4	4.7±0.8
Se 9	0.9±0.3	1.0±0.3	0.7±0.3	0	1.1±0.4
Se 11	0.6±0.2	0.6±0.2	0.6±0.2	0.2	1.9±0.5
Esch 3	0	1.6±0.5	4.6±0.7	3.1±0.6	5.9±0.8
Esch 9	0	0.9±0.3	1.6±0.4	0	1.0±0.3
Esch 11	0.7±0.3	0.5±0.2	1.0±0.3	0	2.3±0.5

3.1.3. Storage for two years. After two years storage, the buds stored with twigs showed extreme sensitivity to the low temperatures. Buds from two (W 3, W 7) of 14 clones survived and differentiated microshoots after storage at only -80 C; the bud explants from the remaining 12 clones did not grow or differentiate microshoots at the five sub-zero temperatures tested. Of the two clones that survived at -80 C, W 7 produced significantly more microshoots (4.1per explant) as compared to W 3 (0.2 per explant).

Based on their regenerative potential under normal conditions , these two clones (W 3, W 7) were classified as 'high shooters', as compared to the remaining 12 clones that are 'low shooters'. It would appear that 'high shooters' are able to survive at -80 C after two years, while the 'low shooters' are unable to do so, when twigs (with dormant buds) are stored at this low temperature. This might suggest that there is a

Table 5. Degree of microshoot differentiation in explants taken from dormant buds stored for five weeks at -80 C, with (+) or without (-) wax-vacuum treatments. (Based on 8-10 explants per treatments)

Clone	-Wax -Vacuum	+Wax -Vacuum	-Wax +Vacuum	+Wax +Vacuum
Esch 2	1.0±0.3	1.7±0.4	1.4±0.4	2.4±0.5
W 29	0.7±0.3	0.9±0.3	1.4±0.4	1.0±0.3

Table 6. Degree of microshoot differentiation in explants taken from dormant buds stored for one year at -80 C, with (+) or without (-) wax-vacuum treatments.

Clone	-Wax -Vacuum	+Wax -Vacuum	-Wax +Vacuum	+Wax +Vacuum
W 29	5.4±0.7	4.0±0.6	3.4±0.6	4.3±0.6
W 80	2.1±0.4	3.6±0.6	2.1±0.4	1.9±0.4
Esch 2	0.9±0.3	1.4±0.4	0.5±0.2	0.9±0.3
Esch 8	2.2±0.5	1.8±0.4	1.5±0.3	2.4±0.5

Table 7. Degree of microshoot differentiation in explants taken from dormant buds stored for two years at -80 C, with (+) or without (-) wax-vacuum treatments.

Clone	-Wax -Vacuum	+Wax -Vacuum	-Wax +Vacuum	+Wax +Vacuum
W 80	0.9±0.3	0.8±0.3	2.3±0.5	1.7±0.4
Esch 8	0.5±0.2	1.7±0.4	1.2±0.3	3.5±0.6

'Darwinian fitness' at -80 C for those clones that are 'high shooters'; those clones which regenerate low number of primary microshoots (less than 3) seem to be at a selective disadvantage at this low temperature for long-term storage.

3.2. Storage of dormant buds

When dormant buds are removed from branches and stored at sub-zero temperatures, they respond somewhat differently to different temperatures over a two year period, as compared to twigs stored with dormant buds.

3.2.1. Storage for five weeks. Only two of 4 clones were tested for their growth and differentiation after five weeks storage at five temperatures.Bud explants from both clones (W 29, Esch 2) differentiated low number of microshoots at 0 C, -5 C, -8 C, and -80 C with or without wax-vacuum treatments. However, bud explants from these two clones did not grow or differentiate microshoots following storage at -18 C. Since the buds remained viable at -80 C after prolonged stoarage, results will only be discussed at -80 C. Within each clone the four treatments involving wax and vacuum, there were no significant differences following storage at -80 C, although in Esch 2 the explants derived from dormant buds coated with wax and stored under vacuum differentiated relatively more microshoots as compared to other treatments (Table 5).

3.2.2. Storage for one year. Bud explants from all four clones did not survive at 0 C, -5 C, -8 C and -18 C after one year of storage, with or without wax-vacuum treatments. On the other hand, bud explants from all four clones (W 29, W 80, Esch 2, Esch 8) stored for one year at -80 C exhibited growth and differentiation of microshoots. Within each clone there were no significant differences between the four treatments (Table 6). However, there were differences in microshoot differentiation between clones and treatments (Table 6).

3.2.3. Storage for two years. Only two clones (W 80, Esch 8) were tested after storage of buds at -80 C for two years. Both showed differentiation of microshoots on bud explants. There were differences between clones and treatments (Table 7). In one clone (W 80) buds stored under vacuum seemed to perform better, while in the second clone (Esch 8) buds coated with wax and stored under vacuum differentiated relatively more microshoots as compared to other treatments (Table 7).

These experiments indicate that buds stored at -80 C may be able to retain their viability, and that low temperatures (-18 C, -8 C, -5 C, 0 C) are not suitable for long-term storage of aspen buds. The storage with wax coating and vacuum may be beneficial for different clones in different ways. These experiments did not indicate that a specific treatment was beneficial to all aspen clones for a long-term storage.

It is apparent from these experiments that dormant buds can be stored under appropriate conditions at sub-zero temperatures. Of the five temperatures tested, -80 C seems to be the optimum temperature for long-term storage of aspen buds. The genotype of a clone and storage conditions seem to affect the storage potential of dormant buds.

REFERENCES

1. Ahuja MR: Somatic cell differentiation and rapid clonal propagation of aspen. Silvae Genet. 32: 131-135, 1983.
2. Ahuja MR: Aspen. In: Handbook of Plant Cell Culture, Vol 4, DA Evans, WR Sharp and PJ Ammirato (Eds.), Macmillan Publishing Company, New York, pp. 626-651, 1986.
3. Ahuja MR: Storage of forest tree germplasm in liquid nitrogen. Silvae Genet. 35: 249-251, 1986.
4. Akihama T and M Omura: Preservation of fruit tree pollen. In: Biotechnology in Agriculture and Forestry, Vol 1, Trees, YPS Bajaj (Ed.), Springer Verlag, Berlin, Heidelberg, pp. 101-112, 1986.
5. Breloh P und G Dieterle: Ergebnisse der Waldschadenserhebung 1985. Allgem. Forstz. 40: 1377-1380, 1985.
6. Herrmann S: Verfahren zur Konservierung und Erhaltung der Befruchtungsfähigkeit von Waldbaupollen über mehrer Jahre. Silvae Genet. 25: 223-229, 1976.
7. Melchior GH, HJ Muhs and BR Stephan: Tactics for the conservation of forest gene resources in the Federal Republic of Germany. Forest Ecology and Management 17: 73-81, 1986.
8. Sakai A: Cryopreservation of germplasm of woody plants. In: Biotechnology in Agriculture and Forestry, Vol 1, Trees, YPS Bajaj (Ed.), Springer Verlag, Berlin, Heidelberg, pp. 113-129, 1986.

SCREENING FOR RESISTANCE TO *HYPOXYLON MAMMATUM* IN *POPULUS TREMULOIDES* CALLUS AND MICROPROPAGATED PLANTLETS

F. VALENTINE, S. BAKER, R. BELANGER, P. MANION and D. GRIFFIN

SUNY College of Environmental Science and Forestry, Syracuse, NY 13210, U. S. A.

Two *in vitro* procedures are being tested for screening trembling aspen for resistance to *Hypoxylon* canker: (1) tissue culture response to fungal culture metabolites and (2) ascospore inoculation of moisture stressed plantlets. In the first procedure, the sensitivity of five clones to the acidic ethyl acetate (AEA) fraction of partially purified culture filtrate was determined by comparing callus growth on ACM medium containing the AEA fraction with that of a control containing non-inoculated culture filtrate. Treatments of 0.1x and 1.0x AEA concentrations were inhibitory compared with controls. The clones varied considerably in callus growth and callus inhibition, with the inhibition relative to growth of the control directly related to that growth. Correlations of the resistance responses of 10 year old ramets to artificial inoculations with mycelial-infested grain show a moderate positive correlation of both callus growth of the control and callus inhibition with canker length and with incidence of branch death. A reasonably strong correlation of lesion diameter in leaf bioassays, a measure of culture filtrate toxicity, and callus inhibition (r=0.74) indicates that host response to the fungal metabolites in these bioassays is similar and perhaps the same. Moreover, both resposes are directly related to the frequency of branch death, which further supports the conclusion of a common factor in these host-pathogen interactions. Host responses to *Hypoxylon* culture metabolites as measured by each of these bioassays have also proved important variables for predicting field resistance of aspen clones by polar ordination and multiple regression analyses, clearly indicating an important role of the toxic metabolites to disease incidence. Procedures for direct inoculation of young aspen with ascospores, the second method being studied, have been developed in assessing the effect of moisture stress on *Hypoxylon* infection of aspen. One to 2 cm plantlets of 6 clones, micropropagated from dormant buds, were stressed by mannitol at various concentrations in ACM medium and inoculated with ascospores on unwounded stems. Mycelial growth appeared after 3 to 4 days, but on the controls and moderately stressed plantlets it remained superficial even after 10 days. In contrast, highly stressed plantlets were invaded by the mycelium and exhibited necrotic lesions at the site of inoculation. The level of moisture stress required for mycelial invasion and lesion development varied (-6 to -12 bars) among the clones. Transplants of the regenerated plantlets will be tested for resistance to *Hypoxylon* and for sensitivity to culture filtrate metabolites.

M. R. Ahuja (ed.), Somatic Cell Genetics of Woody Plants, 181.

INCREASED HERBICIDE TOLERANCE OF *IN VITRO* SELECTED HYBRID POPLAR

C.H. Michler and B.E. Haissig. USDA Forest Service, North Central Forest Experiment Station, Forestry Sciences Laboratory, P.O. Box 898 Rhinelander, WI, 54501, USA.

ABSTRACT

Hybrid poplars, selected for optimal yield in the northern USA and Canada, do not tolerate chemical stresses. Synthetic chemical herbicides are important in intensive silviculture, however poplars are often more sensitive than major weeds to broad spectrum herbicides applied during the growing season. Through our research we have increased the tolerance of four hybrid poplar clones to glyphosate (Roundup[1], Monsanto Chemical Co.) and sulfometuron methyl (Oust, Du Pont Chemical Co.). Leaf explants were subjected to herbicide selection on optimized shoot regeneration medium supplemented with glyphosate or sulfometuron methyl. Regenerated adventitious shoots were rechallenged at the lethal dose of herbicide for non-selected *in vitro* shoots. We were thus able to select shoots with increased tolerance to herbicide-containing tissue culture medium. Initial growth chamber testing of rooted herbicide tolerant shoots confirmed tolerance to glyphosate demonstrated *in vitro*.

1. INTRODUCTION

Herbicide tolerant forest trees are needed to increase survival and productivity of managed forests which may be reduced by weed competition (Nelson et al., 1984a). Herbicides are often more efficient than mechanical cultivation, but cause chemical stress to the tree (Brynes et al., 1978). Broad-spectrum herbicides currently used in forestry may kill woody species as well as the target weeds when applied to actively growing plants (Akinyemiju and Dickmann, 1982). Suboptimal weed control is acceptable where the desired species must not be damaged, but results in less than maximum yield from forest trees. Herbicide tolerant forest trees would thus be an important new crop where weed competition deters productivity (Nelson and Haissig, 1986).

Tolerance to many broad spectrum herbicides is not known to exist in angiospermous tree species and would thus be difficult to achieve through traditional breeding (Nelson and Haissig, 1986). However, herbicide tolerance has been inadvertently selected for in several weed species by continual herbicide application to agronomic crops. Herbicide tolerance may arise due to spontaneous mutations (Bandeen et al., 1982). Horticultural and forestry crops are not subjected to continual herbicide stress over many years, so natural selection would be much more difficult to obtain through the same method. Long sexual generation times for trees

---

[1] Mention of tradenames does not constitute an endorsement by the United States Department of Agriculture or the Forest Service.

M. R. Ahuja (ed.), Somatic Cell Genetics of Woody Plants, 183–189.

preclude breeding as a fast method for producing herbicide tolerant trees (Nelson et al., 1984b), assuming the existence of herbicide tolerant parents.

Two methods may quickly achieve herbicide tolerance in woody plants: direct gene insertion and somaclonal selection. For example, insertion of a bacterial glyphosate tolerance gene into poplar by using *Agrobacterium* increased herbicide tolerance (Fillatti et al., 1987). *In vitro* selection has been used previously to select a small number of plant mutants with increased herbicide tolerance (Chaleff, 1980; Flick, 1983; Shaner et al., 1985). Somaclonal technology allows the study of possible pre-existing or *in vitro* culture induced genetic variation. These variations may be expressed, described, and tested from a single passage in the tissue culture environment.

For the herbicides used in our research, glyphosate and sulfometuron methyl, the modes of action are known (Amrhein et al, 1983; Chaleff and Mauvais, 1984) and tolerance to the herbicides has resulted from mutation at a single gene locus (Chaleff and Ray, 1984; Comai et al., 1985). Both of these attributes allow for *in vitro* selection as opposed to conventional breeding because only a single mutation may have to be selected to achieve tolerance. Tolerance mechanisms due to difference in conformation or production can be tested by enzyme activity analysis.

The purpose of our research was to determine if somaclonal selection could be used to obtain herbicide tolerances in poplars. Results indicated that large amounts of tolerance to either glyphosate or sulfometuron methyl were obtained in several clones of hybrid poplars.

2. MATERIALS AND METHODS

2.1 Plant Materials

The following four hybrid poplar clones were used:

Clone[1]	Parentage	Section
NC-5339	*P. alba* × *P. grandidentata*	Leuce
NC-5272	*P. nigra* × *P. laurifolia*	Aegeiros-Tacamahaca
NC-11390	*P. maximowiczii* × *P. trichocarpa*	Tacamahaca
NC-5331	*P. nigra betulifolia* × *P. trichocarpa*	Aegeiros-Tacamahaca

[1] North Central Forest Experiment Station accession numbers.

2.2 Establishment of Continuous Shoot Cultures

Shoot tips (2 cm) were severed from greenhouse stock plants from cuttings of immature trees, placed into sterile water, stirred for 10 m in dilute commercial detergent, sterilized in 10% Clorox for 10 m, and rinsed three times in sterile purified water. The shoot tips were placed horizontally in GA-7 culture vessels on basal Murashige and Skoog (MS) medium (1962) containing 100 mg/l myo-inositol, 200 mg/l casein hydrolysate, 2% sucrose, 0.1 mg/l benzyladenine, and 0.6% Difco Bacto agar (pH 5.8). The shoot cultures were placed under an 18 h photoperiod in cool white light (30 $uE/m^2/sec^1$) at 26°C in a controlled environment room. Shoot cultures were subcultured every four weeks on the identical medium. Adventitious and axillary shoots were transferred by placing a 2 to 3 cm shoot tip vertically or horizontally onto the medium or by transfer of the entire multiple shoot clump. These continuous shoot cultures were the starting plant material used in all tests.

2.3 Optimization of Adventitious Shoot Production

Leaf pieces (apical and basal sections removed) of the four hybrid clones were placed in factorial experiments with the plant growth regulators benzyladenine (BA) and naphthaleneacetic acid (NAA). A 5x5 factorial arrangement of growth regulators was used with concentrations of both at 0, 0.01, 0.1, 1 and 5 mg/l. Explants were placed in disposable quad petri dishes; the plates were wrapped with parafilm. Four replications with two samples each were placed in a randomized complete block design. The cultures were grown under the same environmental conditions as described above for the shoot cultures. Shoot production data were taken after six weeks.

2.4 Somaclonal Variant Selection

Leaf pieces were placed on shoot regeneration medium (Table 1) containing herbicide for 30 days in the dark. Selections were made on either 0, 100, 250, 500, or 1000 umol/l glyphosate or 0, 27, 69, 137, 206, and 274 nmol/l sulfometuron methyl. After the initial selection for 30 days, the leaf pieces were either allowed to remain on herbicide selection medium or were transferred to shoot regeneration medium with 0.1 mg/l BA as the only growth regulator. Leaf pieces were continually subcultured at 30-day intervals in the same medium until shoots >1 cm long had been regenerated. Shoots that regenerated on medium that always contained herbicide were subcultured for three consecutive 30-day periods on MS + 0.1 mg/l BA medium. Shoots >3 cm long were rooted in Ekco foil trays and transferred to the growth chamber.

2.5 Determination of Herbicide Lethality

Shoot tips (2 cm) from continuous shoot cultures were vertically explanted in petri plates of MS containing 0.1 mg/l BA + 30 g/l sucrose (solidified with 0.6% Difco Bacto agar). Filter-sterilized glyphosate (used as Roundup) was added before solidification at final concentrations of 0, 5, 10, 25, 50, 75, 100, 250, 500, and 1000 umol/l; sulfometuron methyl was added in the following amounts: 0, 3, 7, 14, 21, 27, 69, 137, 206, and 274 nmol/l. The shoot cultures were grown in the light for four weeks and analyzed for necrosis. The experiment was replicated twice with four samples each per clone and treatment.

2.6 Rechallenge of Putative Herbicide Tolerant Somaclones

Shoots that regenerated on medium that contained herbicide for only the initial 30 days were rechallenged for three consecutive 30-day intervals on medium with a toxic concentration (to non-selected plants) of the herbicide (Table 2). Survivors were subcultured on MS + 0.1 mg/l BA medium without herbicide for three consecutive 30-day intervals. Subsequently, their shoots were rooted to produce plants for greenhouse testing.

2.7 Preliminary Ex Vitro Testing of Somaclonal Variants

Stock plants were established for each of the somaclonal variants in the greenhouse. The stock plants were pruned to induce formation of lateral branches and softwood cuttings (5 cm) severed from these plants were dipped in 1000 ppm NP-IBA (N-phenyl indolyl-3-butyramide) and rooted under mist in commercial soil-less peat-vermiculite mix. In initial tests, only glyphosate tolerant selections of NC-5339 and NC-5272 were sprayed along with control rooted cuttings of tissue culture produced stock plants of these clones. Roundup was applied to the softwood cuttings as a solution of 2.125 mmol/l based on a rate of 0.14 kg/ha. Herbicide tolerant plants were compared to non-selected plants regenerated in tissue culture for differences in leaf and stem morphology.

3. RESULTS

Our results indicate that shoot regeneration (Table 1) using leaf pieces was greatest for NC-11390 followed by NC-5339 and the other two clones (NC-5272 and NC-5331). The optimal growth regulator combination was different for three out of four clones. The highest cytokinin/auxin ratio resulted in the greatest number of adventitious shoots regenerated as demonstrated with NC-11390.

Table 1. Growth regulator combinations that resulted in the most adventitious shoot regeneration from hybrid poplar leaf explants.

Clone	Growth Regulator Treatment (mg/l)		Shoot Number[a]
	BA	NAA	
NC-5339	1.0	1.0	16.1 + 1.1b
NC-5331	1.0	1.0	2.5 + 0.4c
NC-11390	1.0	0.1	29.5 + 3.4a
NC-5272	0.1	0.1	3.5 + 0.6c

[a] values in a column followed by a different letter are significantly different at the 5% level using mean separation by LSD.

LD_{100} concentrations for the four clones were determined for both herbicides (Table 2). These levels of herbicide were used to rechallenge regenerated shoots from herbicide selection experiments. With glyphosate-amended medium, NC-5331 microshoots had the greatest natural tolerance (250 umol/l) and NC-5339 had the least at 50 umol/l. With sulfometuron methyl as the selective agent, NC-5272 was the most sensitive (3 nmol/l) and NC-5339 was the least sensitive (137 nmol/l).

Table 2. Herbicide concentrations (LD_{100})[a] used to rechallenge putative herbicide tolerant somaclonal variants of hybrid poplars.

Clones	Glyphosate (umol/l)	Sulfometuron methyl (umol/l)
NC-5339	50	.137
NC-5331	250	.075
NC-5272	75	.003
NC-11390	75	.015

[a] LD=Minimum dosage at which 100% of plants were killed by herbicide in agar medium for vertically positioned shoot tips.

Following rechallenge with glyphosate, the largest number of putative tolerant shoots remained for hybrid clone NC-5339, the fewest for NC-5331 (Table 3). Despite the lowest number of tolerant microplants in NC-5272 and NC-5331, these clones were the most efficient producers of tolerant shoots, 4.2% and 4.3%, respectively (Table 3).

Table 3. Frequency of regeneration of herbicide tolerant hybrid poplar shoots[a] following rechallenge with toxic concentrations of glyphosate and sulfometuron methyl.

	Frequency of Herbicide Tolerant Shoots Regenerated			
	Glyphosate		Sulfometuron Methyl	
Hybrid CLone	No.	(%)	No.	(%)
NC-5339	62/2581[b]	2.4	3/2322	0.13
NC-11390	36/4720	0.76	4/4248	0.094
NC-5272	26/620	4.2	36/504	7.1
NC-5331	17/400	4.2	2/360	0.56

[a] Table 2 lists herbicide concentrations used in rechallenge of cultures.

[b] Number of tolerant shoots regenerated from leaf explants/predicted number of shoots, those that would have regenerated without selection pressure.

Following rechallenge with sulfometuron methyl, NC-5272 had the greatest number of remaining tolerant shoots (36) and was the most efficient producer (7.1%, Table 3). The other three clones produced only a few shoots; NC-11390 was the least efficient with four live shoots of 4248 possible regenerants.

Glyphosate tolerant plants had normal leaf morphology, but selections from all clones except NC-5339 from sulfometuron methyl testing had altered leaf morphology resulting in wavy leaf margins with variant leaf serrations. This altered leaf morphology has been stable for four months following plant establishment in the greenhouse. Long term morphology is unknown.

Preliminary spray testing of putative glyphosate tolerant variants of NC-5339 and NC-5272 confirmed their increased tolerance to the herbicide. Two of eleven NC-5339 variants and three of six NC-5272 variants survived after having been sprayed with Roundup (0.14 kg/ha) along with micropropagated control plants. These plants continued shoot growth at the same rate as unsprayed control plants. The foregoing tests are only now being replicated so results reported here are provisional. We are also conducting greenhouse tests of other putative herbicide tolerant clones. The remaining herbicide tolerant selections are also undergoing greenhouse testing to determine if increased tolerance is stable *ex vitro*.

4. DISCUSSION

We were successful in regenerating adventitious shoots from all four hybrid poplar clones with increased tolerance to the herbicides *in vitro*.

This was proven by determining LD_{100} levels of herbicide to use as selection pressure with each plant challenge in vitro. Normally non-tolerant tissue cultures died at the lethal dosage, either at initial selection or at rechallenge with herbicide at the cellular, organ, or whole plant level allowing selection of tolerant tissue cultures at each stage of development.

There was a clonal difference in the propensity to regenerate tolerant microshoots. The percentage of herbicide tolerant shoot regenerants from the Aegeiros-Tacamahaca intersectional hybrids (NC-5331 and NC-5272) was higher than that from Leuce (NC-5339) and Tacamahaca (NC-11390) intersectional hybrids, but the sampling size within sections may not have been sufficient to draw firm conclusions. Clones with the lowest frequency of shoots regenerated also produced a higher percentage of tolerant microshoots. Differences in frequency of somaclonal variant production may be reduced if the regeneration medium is further refined.

Initial testing of rooted softwood cuttings from putative herbicide tolerant plants showed an increased tolerance to glyphosate that is greater than non-selected tissue culture produced plants. The tolerance is stable after the plants have been selected and rechallenged in vitro.

Results from our tests indicate that some hybrid poplars are amenable to somaclonal selection for increased herbicide tolerance. Future studies will determine biochemical or whole plant mechanisms for herbicide tolerance and levels of tolerance in field-grown plants.

5. ACKNOWLEDGMENTS

We greatly appreciated the technical assistance of Mr. Edmund Bauer and Ms. Therese Voelker and the clerical assistance of Ms. Edith Franson. This research was performed under Interagency Agreement Number DE-A105-800R20763 for the US Department of Energy.

6. LITERATURE CITED

Akinyemiju, O.A. and D.I. Dickmann. 1982. Variation among 21 Populus clones in tolerance to simazine and diuron. Can. J. For. Res. 12:708-712.

Amrhein, N., D. Johanning, J. Schab, and A. Schulz. 1983. Biochemical basis for glyphosate-tolerance in a bacterium and a plant tissue culture. FEBS Lett. 157:191-196.

Bandeen, J.D., G.R. Stephenson, and E.R. Cowett. 1982. Discovery and distribution of herbicide-resistant weeds in North America. In: Herbicide Resistance in Plants. H.M. Lebaron and J. Gressel, eds. pp. 10-19. John Wiley and Sons, NY.

Byrnes, W.R., C. Merritt, and J.M. Braun. 1978. Growth response of black walnut and yellow-poplar to site improvement and weed control. In: Proc. Central Hardwood For. Conf. P.E. Pope, ed. pp. 412-424. Purdue University.

Chaleff, R.S. 1980. Further characterization of pichloram tolerant mutants of Nicotiana tabacum. Theor. Appl. Genet. 58:91-95.

Chaleff, R.S. and C.J. Mauvais. 1984. Acetolactate synthase is the site of action of two sulfonylurea herbicides in higher plants. Science 224:1443-1445.

Chaleff, R.S. and T.B. Ray. 1984. Herbicide-resistant mutants from tobacco cell cultures. Science 223:1148-1151.

Comai, L., D. Facciotti, W.R. Hiatt, G. Thompson, R.E. Rose, and D.M. Stalker. 1985. Expression in plants of a mutant aroA gene from Salmonella typhimurium confers tolerance to glyphosate. Nature 317:741-744.

Fillatti, J.J., J. Sellmer, B. McCown, B. Haissig, and L. Comai. 1987. Agrobacterium mediated transformation and regeneration of Populus. Mol. Gen. Genet. 206:192-199.

Flick, C.E. 1983. Isolation of mutants from cell culture. In: Handbook of Plant Cell Culture. D.A. Evans, W.R. Sharp, P.V. Ammirato, and Y. Yamada, eds. Vol. 1, pp. 442-460. Macmillan Co., NY.

Murashige, T. and F. Skoog. 1962. A revised medium for rapid growth and bioassays with tobacco tissue cultures. Physiol. Plant. 15:473-497.

Nelson, L.R., D.H. Gerstad, and P.J. Minogue. 1984a. Use of herbicides for industrial forest vegetation management in the southern United States. In: Chemical and Biological Controls in Forestry. W.Y. Garner and J. Harvey, Jr., eds. ACS Series 238:11-23. American Chemical Society, Washington, D.C.

Nelson, N.D., B.E. Haissig, and D.E. Riemenschneider. 1984b. Applying the new somaclonal technology to forestry. In: Proc. TAPPI 1984 Res. and Dev. Conf. pp. 27-34. Appleton, WI. TAPPI, Atlanta, GA.

Nelson, N.D. and B.E. Haissig. 1986. Biotechnology of herbicide resistance for enhanced forest productivity. In: Stress Physiology and Forest Productivity. T.C. Hennessey, P.M. Dougherty, S.V. Kossuth, and J.D. Johnson, eds. pp. 191-215. Martinus-Nijhoff, Dordrecht.

Shaner, D., T. Malefyt, and P. Anderson. 1985. Herbicide-resistant maise through cell culture selection. In: Biotechnology and Its Application to Agriculture. L.G. Copping and P. Rodgers, eds. BCPC Monograph 36. pp. 45-50. British Crop Protection Council.

THE OCCURRENCE OF ENDOGENOUS BACTERIA IN BIRCH SHOOT CULTURES

AMANDA M GORDON AND IAN R BROWN
DEPARTMENT OF FORESTRY, UNIVERSITY OF ABERDEEN, ST MACHAR DRIVE,
ABERDEEN AB9 2UU, SCOTLAND

INTRODUCTION

The production of clonal material by micropropagation is becoming increasingly important to both the horticultural and forestry industries. One of the factors to be taken into consideration is that of bacterial contamination of stock cultures. Until recently this was believed to be principally due to the presence of exogenous bacteria. These contaminants would either be present due to an inadequate initial sterilization procedure or introduced during subsequent culture operations. The presence of endogenous bacteria in tissue cultures has been reported by Debergh (1, 2, 3). This paper describes the isolation and identification of an endogenous bacterium from shoot cultures of *Betula pendula*, *Betula pubescens* and *Betula nana*.

PROCEDURE

Materials and Methods

Seeds from a selection of known *Betula pendula* and *Betula pubescens* genotypes were sterilised, germinated and micropropagated. The shoot cultures were tested for bacterial contamination by excising a piece of tissue and smearing its cut edge over the surface of a potato dextrose agar (PDA) plate. The plates were incubated at a temperature of 25°C for 72 hours. PDA plates were flooded with a bacterial suspension and multodiscs were applied to determine the most effective antibiotic.

RESULTS

The effect of a number of antibiotics on the growth of the bacterium are listed in Table 1. Streptomycin and tetracycline were the only two drugs capable of inhibiting growth and of these tetracycline was the more efficient. Consequently, tetracycline was added to the basic culture medium at a concentration of 10mg l^{-1}. The explants were cultured in the presence of the antibiotic for 14 days and then transferred to fresh culture medium. This resulted in an inhibition in bacterial growth for up to 16 weeks. After this period the number of contaminated cultures gradually increased (Table 2). This reflects the bacteriostatic rather than the bacteriocidal properties of tetracycline. Subsequently, rifampicin, a bacteriocide, was tested for its effect on the growth of the bacterium. At a concentration of 20mg l^{-1} rifampicin totally inhibited bacterial growth and its inclusion in the culture media results in the decontamination of cultures which are grossly infected with endogenous bacterium. In contract with tetracycline, rifampicin has no phytotoxic effects on the shoot cultures.

To date identification of the bacterial isolate has been frustrated by its extremely specific growth requirements. It has not been possible to sustain its growth on a number of the required test media and this has led to problems in identifying it beyond the genus level. The organism has been identified as a strain of a *Bacillus* species from spore group II. It is a gram negative rod with an optimum growth temperature and pH of 25°C and 5.5 respectively.

M. R. Ahuja (ed.), Somatic Cell Genetics of Woody Plants, 191–193.

Table 1: THE EFFECT OF ANTIBIOTICS ON ENDOGENOUS BACTERIAL GROWTH

A suspension of the bacterium was plated onto potato dextrose agar plates and multodiscs impregnated with antibiotics were applied.

ANTIBIOTIC	EFFECT ON GROWTH
Chloramphenicol	-
Cephaloridine	-
Erythromycin	-
Kanamycin	-
Penicillin	-
Ampicillin	-
Streptomycin	+
Tetracycline	++

+ = Inhibition - = No Inhibition

Table 2: THE EFFECT OF TETRACYCLINE ON BACTERIAL GROWTH

Treated cultures were grown in medium supplemented with tetracycline (10mg l^{-1}) for 14 days. They were then transferred into antibiotic-free medium. Untreated cultures were also transferred into fresh medium after 14 days.

TIME IN CULTURE (WEEKS)	INFECTED CULTURES (%)	
	TREATED	UNTREATED
4	5	12
8	9	50
12	11	55
16	12	60
20	30	65

DISCUSSION

The existence of endogenous bacteria in tissue cultures of woody plant species is becoming increasingly recognized. However, the detection of these organisms is complicated by their ability to remain dormant in spore-form for a number of months. From our experience the bacteria become evident when the culture is undergoing a period of stress or slow growth ie. immediately after sub-culture. It is at this stage that the growth of the bacteria can accelerate to a rate which is deleterious to further development. This results in the browning and eventual necrosis of the culture. The causal factors are unknown but it is possible that the bacteria reside in the plant vascular system and their sudden increase in mass results in a blockage of the vascular tissues. Alternatively, the bacteria may produce a metabolic by-product which in high concentrations is toxic to the plant tissues. These endogenous bacteria do not appear to be pathogenic in the field as infected plantlets which are rooted and removed from *in vitro* conditions grow and continue to flourish.

The pathogenecity of endogenous bacteria is apparently related to the artificial growth conditions of *in vitro* culture. At present the inclusion of antibiotics in the plant growth medium is the only means of controlling their growth. An alternative may be the development of culture media and techniques which would erradicate the stress factors responsible for the initial germination of the dormant bacterial spores.

REFERENCES

1. Debergh, P C and Maene, L J. (1984) Pathological and physiological problems related to the in vitro culture of plants. Parasitica 40: 69-75.
2. Debergh, P C. (1986a) Micropropagation of herbaceous plants. In: Micropropagation in horticulture - practice and commercial problems. Symp. Nottingham (UK). Ed P Alderson, in press.
3. Debergh, P C. (1986b) Recent trends in the application of tissue culture of ornamentals. In: Proceedings of the VIth International Congress Plant Tissue and Cell Culture. Minnesota (USA), in press.

ESTABLISHMENT OF A STERILE FOAM CULTIVATION SYSTEM: ELIMINATION OF FORMALDEHYDE

K. GEBHARDT

Forschungsinstitut für schnellwachsende Baumarten
Prof. Oelkers Str. 6, D-3510 Hann. Münden, F.R.G.

ABSTRACT

The effective elimination of formaldehyde is a prerequisite for the routine use of phenol resin foam in micropropagation. Formaldehyde which has been freed from the foam substrate during autoclaving can be eliminated by the use of ammonia solution. For routine use, a pretreatment of foam with 10 ml 2 normal NH_4OH per 70 cm^3 of foam volume can be generally recommended. The pretreatment effect is tested by the use of Schiff's reagent. Pretreated foam can be used in a foam cultivation system for rooting of various shoot tip cultures. As demonstrated with shoot tip cultures from sour cherries (*Prunus cerasus*) it will also support rootless cultures, and it provides an excellent transport system.

1. INTRODUCTION

As reported in Plant Science Letters (1), foam substrates can be used as support for rooting of shoot tip cultures instead of an agar medium. Foam substrates are manufactured from different resins, foaming agents and catalysts. They are used in horticulture as a substitute for colloid substrates such as peat, vermiculite or rock wool in non sterile cultivation systems. If supplied with a liquid nutrient solution, the capillarity system of the foam provides the nutrient supply for any plant material at the foam surface. Foam substrates differ largely in volume weight, pore size, colour, heat stability and water retention. In the case of heat treatment, foam substrates may loose structure and hydrophobicity as well as release toxic compounds. Wheeler and associates at the University of Wisconsin (5) described the toxicity of polyurethane plugs to plant seedlings, and attributed the toxicity to residual tertiary amines or isocyanate compounds. In early experiments, we noticed the release of formaldehyd from a phenol resin based foam upon autoclaving in a culture tube. Several attempts to avoid the formaldehyde release from the foam substrate were tried, but no treatment of the foam prior to autoclaving proved to be effective. This is due to the fact that formaldehyde can be freed from newly opend pores during the course of autoclaving. Our report here concerns the elimination of formaldehyde and the effective use of a phenol resin foam in a sterile cultivation system for shoot tip cultures in general.

M. R. Ahuja (ed.), Somatic Cell Genetics of Woody Plants, 195–199.

2. PROCEDURE

2.1. Equipment and materials

- foam delivered by
  AGRIMEDIA; Rudolf Diesel Str. 8, D-6719 EISENBERG, F.R.G.
- autoclavable dispensor (8 ml graduation) for nutrient solution;
- 0.5 ml pipettor/syringe;
- autoclavable culture tubes or other vessels:
  prefer test tubes (25 x 150 mm) made of glass with a slim formed edge, use glass caps 28 x 40 mm to close test tubes;,
- Solutions:

A. Aqueous ammonia (NH_4OH) at concentrations from 1 - 3 Normal.
   A 2 normal (2 N NH_4) solution is prepared from 25 % ammonia stock solution by dissolving 30 ml in 2 liter (3 ½ pints) distilled water. Make up to 100 ml of 1 to 3 N NH_4OH with distilled water. Store tightly closed.

B. "Schiff's" reagent, Fluka No. 84655, Fluka AG, D-7910 Neu-Ulm,

C. Liquid medium, prepared according to the specifics.

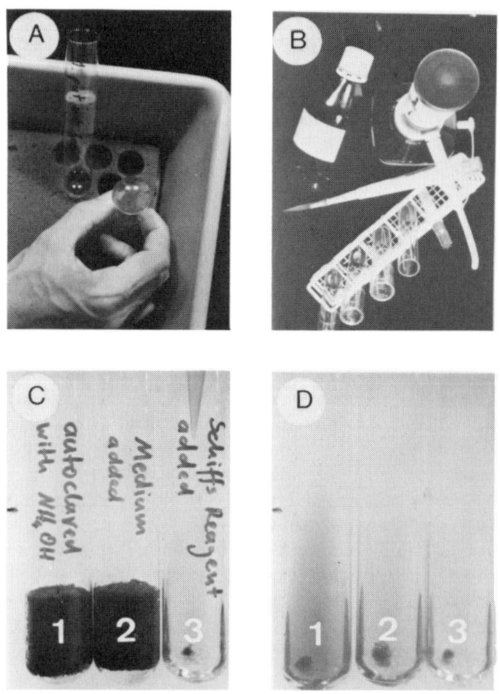

FIGURE 1: Procedure.

A. Glass tubes are used to remove foam cylinders.

B. Autoclavable dispensor, micropipettor, test tubes and Schiff's reagent.

C. Steps in formaldehyde test procedure from left to right:
1. foam cylinder autoclaved with 2 N NH_4OH
2. Sterilized nutrient solution (8 ml) added and incubated for 4 hours.
3. Foam cylinder removed and Schiffs reagent added.

D. Discoloration from left to right:
1. control, without ammonium pretreatment exhibits strong discoloration;
2. slight discoloration with a 0.5 N ammonia solution used in pretreatment procedure;
3. no discoloration with a pretreatment using 1.0 N ammonia solution

2.2. Preparation of liquid nutrient solution

2.2.1 Dissolve and combine each component, bring up to final volume and desired pH.

2.2.2 Autoclave solution in glass jar closed with an autoclavable dispensor (1.1 kg/cm^2 pressure, 121°C for 15 min.)

2.3. Preparation of the foam cultivation system and foam pretreatment.

2.3.1 Use test tubes (2.5 x 15 cm) in order to remove foam cylinders (diameter: 2.5 cm; height 3.0 cm; vol.: 14.72 cm^3) as demonstrated in Fig. 1

2.3.2 Push down to the bottom of the test tube.

2.3.3 Add 2 ml ammonia solution (NH_4OH) at a range of concentrations from 1 to 3 normal. Repeat this procedure in triplicate.

2.3.4 Close test tubes and autoclave with the additions of the ammonia solution for 17 min. at 123 °C. Remove steam before it condenses.

2.4. Evaluation of effective formaldehyde elimination

2.4.1 Add 8 ml of liquid nutrient solution at room temperature to autoclaved test tubes of interest.

2.4.2 Allow incubation of nutrient solution and foam cylinder for at least 4 hours.

2.4.3 Remove foam cylinder and add 0.5 ml of Schiff's-Reagent to 2.5 ml of the remaining nutrient solution.

2.4.4 Evaluate discoloration (violet) after 2 hours. Extinction of solution at 544 nm wavelength is proportional to the formaldehyde concentration. No discolouration should be visible at high concentrations of ammonia pretreatment.

2.5. Evaluation of pH-shift

2.5.1 Determine pH of autoclaved nutrient solution.

2.5.2 Add 8 ml of liquid nutrient solution at room temperature to autoclaved test tubes of interest.

2.5.3 Allow incubation of nutrient solution and foam cylinder for 4 hours.

2.5.4 Remove foam cylinder, press out nutrient solution and determine pH.

3. RESULTS

Phenol resin foam is delivered onto plates 3 cm in length and can be used in a foam cultivation system as described in Fig. 1 (preparation, handling) and Fig. 2 (application). The interior surface of the system can be wet sterilized in an autoclave; however, depending upon deviations in the manufacturing process and time of storage, varying amounts of formaldehyde will be freed from the foam during autoclaving. Nutrient solution became contaminated if it was heated in presence of formaldehyde releasing foam. Schiff's reagent added to contaminated nutrient solution formed a coloured (violet) product with formaldehyde, thus making contamination obvious. Formaldehyde is easily soluble in water and can be retained by the foam. A pretreatment of the foam with ammonia solutions (0.5 - 2 normal NH_4OH) proved to be useful in the elimination of formaldehyde , as is shown in Table 1 . As described in "Römpp's Chemie Lexikon" (p. 1180), formaldehyde dissolved in water can be transferred to hexamethylenetetramine $[(CH_2)_6 N_4]$ via formaldimine if it is heated together with ammonia solution. Hexamethylenetetramine is a salt (mol. weight 140.19) which is very soluble in aqueous solutions. It reacts chemically with anorganic salts present in nutrient solutions (Beilstein E 3/4, p. 1680). Solutions of hexamethylenetetramine have been used in citriculture (Römpp's Chemie Lexikon, p. 1467) because of its fungizide effect. The reverse reaction from hexamethylenetetramine to formaldehyde was observed in pure distilled water, however not in nutrient solutions following 4 weeks of incubation. Ammonia dissolved in water evaporates when heated and can be removed by gas exchange during autoclaving or thereafter. The foam cultivation system is completely established when autoclaved nutrient solution is poured on formaldehyde-freed foam in a sterile manner. Effective elimination of ammonia avoids a shift in the pH of nutrient solutions used in the foam cultivation system. The pH of media is generally adjusted prior autoclaving. According to medium composition and addition of buffering substances, the pH may raise more than one degree after autoclaving.

Table 1: Elimination of Formaldehyde

Pretreatment of foam cylinders	Discoloration of Nutrient solution Extinktion:544 nm Test with Schiffs R.	pH-shift of nutrient sol. on pretreated foam cylinder
1. control, foam not autoclaved	0.018	--
2. foam autoclaved with 2 ml H_2O	0.580	+ 0.03
3. 2 ml 0.5 N NH_4OH	0.188	+ 0.12
4. 2 ml 1.0 N NH_4OH	0.030	+ 0.37
5. 2 ml 1.5 N NH_4OH	0.036	+ 0.40
6. 2 ml 2.0 N NH_4OH	0.028	+ 0.56

4. APPLICATION

4.1 Cultivation of shoot tip cultures with pretreated foam:

4.1.1 Rooting of shoot tip cultures of *Prunus cerasus*.

Optimum rooting was acheived with elongated shoots from shoot tip cultures of sour cherry varieties (Fig. 2 A and B), using 8 ml of the liquid nutrient solution (a modified Murashige & Skoog high salt medium) in the foam cultivation system described. The liquid medium was of the same composition , including 1 ppm indolebutyric acid, as used for rooting in an agar solidified medium (2). A relation of 8 ml nutrient solution to a 14.7 cm^3 foam cylinder provided roots with oxygen and sufficient nutrients.

FIG.2

A. Sour cherries rooted in agarous medium and in foam substrate (left).
B. Sour cherry (*Prunus cerasus* cv. Köröser), exhibiting a branched root system in foam substrate

Roots penetrated the foam substrate within a 4 week subculture period. If roots become visible the plantlets will be ready for transfer to nonsterile conditions. Rooted plantlets are removed inclusively the foam cylinder from test tubes. Nutrient solution is easily washed off the foam by tap water. If plantlets rooted in foam are dipped in a fungizide solution the foam will keep the fungizide solution after potting in a wet mixture of peat and sand. Plantlets should be kept at 100 % humidity (intermittend mist or covered by transparent plastic-foil) until roots have penetrated the soil substrate.

FIGURE 3:

Root hairs indicating oxygen in the root microenvironment.

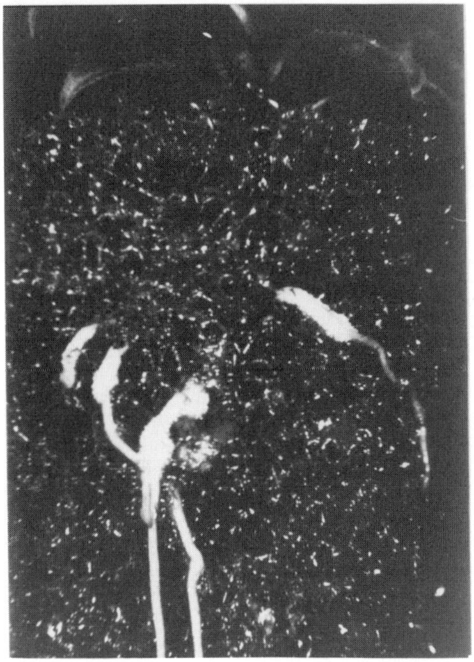

4.1.2 Transport of rooted shoots in a foam cultivation system:

Rooted plantlets can be kept and transported in the sterile foam cultivation system for a few month at very low light level and below room temperature. In contrast to plantlets rooted in agarous media they will not suffer so hard from oxygen deficiency. The presence of oxygen in the root micro-environment is indicated by the development of root hairs (see Fig. 3). We have stored plantlets at about 20 $\mu E/m^2$ sec. and 18°C for more than 4 month. The nutrient solution was changed only one time during storage.

REFERENCES

1. Gebhardt K: Development of a sterile foam cultivation system in vitro. Plant Science Letters 39, 141-148, 1985a

2. Gebhardt K: Self-rooted sour cherries in vitro: auxin effects on rooting and isoperoxidases. Acta Horticulturae 169, 341-349, 1985b.

3. Murashige T and Skoog F: A revised medium for rapid growth and bioassays with tobacco tissue cultures. Physiol. Plant. 15, 473-479, 1962.

4. Römpps Chemie Lexikon 3, Franckh'sche Verlagsbuchhandlung, Stuttgart, 1973.

5. Wheeler R, Schwarzkopf S, Tibbitts T and Langhans R: Elimination of toxicity from polyurethane foam plugs used for plant cultures. Hort.Science 20, 448-449, 1985

TESTING AND DEPLOYMENT OF BRAVE NEW PLANTLINGS

W.J. Libby
Department of Genetics and of Forestry & Resource Management
University of California, Berkeley 94720, USA

0. ABSTRACT

Forestry is a conservative profession, and rightly so. This paper addresses the field testing and deployment of whole trees resulting from genetic manipulation of somatic cells. It particularly focuses on those requirements of field testing that have implications for the choice of plant materials used for <u>in-vitro</u> cell culture and manipulation. Responsible field testing requires long periods of time, and it evaluates the total performance of a tree or clone. Widespread deployment can responsibly commence only following successful field testing. Field testing will be more efficient and deployment can commence sooner if the genetically engineered clones come from well known clones, or from well known families, or from well tested families, or at least from trees of well known provenance. There is some concern regarding the maintenance and support of a professional infrastructure for field testing genetically novel trees.

1. INTRODUCTION

I am here as something of an outsider, and surely don't intend to tell you how to select genotypes or how to conduct experiments during the fundamental and developmental phases of research in forest biotechnology. Rather, it is my job to anticipate your successful research, and to look forward to the deployment of the products of that research in production forestry. The successes and progress reported in several of the papers at this meeting indicate that the time for such considerations has come. I won't provide details of the theory and practice of field testing and production-level deployment. For those interested, these are provided in (13). Instead, I will offer some of the reasoning behind those details of practice and theory, and will make some suggestions as to the kinds and amounts of genetic material that seem appropriate as we move toward field testing and deployment.

2. PROBLEMS WITH DEPLOYMENT

2.1. <u>Forestry is a conservative profession, and rightly so</u>

While I often join many of you in being frustrated at the forestry profession's reticence to embrace new ideas, and its slowness to adopt new procedures that appear better, there are some defensible reasons for such attitudes. Unlike most other economically important organisms, trees remain present and alive in one place for the full 12 months of every year, and they do this for several years, for decades, for centuries and even for millenia. During these long unbroken periods, the successful trees must be able to endure every adverse physical and biotic component of their environments, including unusual events that may occur cyclically or episodically only a few times per century. To implement the widespread deployment of novel trees in a region (whether the novelty be associated with a non-native evolutionary history or with biotechnological manipulation) is to risk widespread failure due to some single adverse event, and is thus courting a regional disaster.

2.2. <u>Conservative foresters demand genetic diversity</u>

It seems likely that most products of forest biotechnology will be deployed as clones, although some will enter sexually reproducing populations. As the forestry

M. R. Ahuja (ed.), Somatic Cell Genetics of Woody Plants, 201–209.

profession begins to deal with clonal forestry, the safety of clonal plantations is becoming a major concern. The theoretical approach, backed by some experimental data, has yielded three general guidelines (5, 6, 7, 8, 9, 10, 11): (a) mixtures of large numbers of clones are about as safe as similar mixtures of genetically diverse seedlings; (b) mixtures or regional deployment of very small numbers of clones are not safe, and commitment to 2-4 clones is often worse than monoclonal deployment; and (c) regional deployment of modest numbers (in the range 7 to 99 clones) is about as safe as deployment of larger numbers of clones and offers substantial additional advantages as well (14).

3. PROBLEMS WITH TESTING

3.1. Time

While it will often be possible to demonstrate quickly that some worthwhile biotechnological manipulation has been accomplished, field testing to a level of reliability that will support responsible widespread deployment is not so quickly done. This is particularly true if the products of biotechnological engineering are something really new and different, since such engineered trees may not have adaptation to a given region's usual and unusual events in their genetic baggage. This "time problem" is very bad news for those who propose radical biotechnology as a shortcut to major genetic advances in forestry. Really novel trees should be stringently tested for long periods of time before they are widely deployed to large-scale or high-value uses. The effects of compound interest on the investments in radical biotechnology and in subsequent testing can make the costs very high before substantial benefits are recovered.

3.2. Total performance

As indicated above, testing in anticipation of substantial deployment must test the total performance of a genetically engineered tree, not just its performance in the few traits that have been modified. This argument has particular force when the new clones are selected somaclonal variants. If the culture of origin was producing a high frequency of somaclonal variation, one must suspect and should test the possibility that cryptic or uncharacterized genetic changes have occurred in other parts of the selected tree's genome. This poses a difficult task, as one doesn't know quite what one is looking for. If sexual propagation follows deployment, some of this variation may survive meiosis and add a kind of genetic load to the population.

3.3. Number and characteristics of test sites

If the population, family or clone performance of a plant is well known prior to its genetic manipulation, then an attractive strategy is to begin with a modest amount of testing in conditions similar to those where the unmodified plants are known to do well. Those candidate plants that do well in their "origin" conditions can then be tested more widely, to determine the extent of their possible deployment. However, if biotechnology produces new plants whose origins can no longer be logically identified, or if such plants are of uncertain origin, then these need an alternative test strategy. If their creation was already expensive, and few such new candidates are produced, then the logical strategy is to clonally test these hopeful monsters over the full range of reasonable interest in the first round of field testing. If large numbers of relatively inexpensive new candidates are produced, then each such clone will probably be first tested on only a few sites, and, if their origins are uncertain, such initial testing will have a hit-or-miss element for interactive clones (13).

3.4. Propagule comparability

That propagules of seed origin (seedlings) and of cutting origin (stecklings) often differ from each other in important respects is well established (16, 17). Many such differences remain statistically, biologically and economically significant after at least seven growing seasons (2), and a recent Australian study of seedling and steckling *Pinus radiata* documents a substantial economic difference at harvest, after 36 years (3). Early results becoming available with species such as *Pinus taeda*, *P. radiata* and *Sequoia sempervirens* indicate that propagules of tissue-culture and/or organ-culture origins differ from both seedlings and stecklings (4, 18, plus a Berkeley *S. sempervirens* test).

Quantitative and qualitative comparisons among such different kinds of propagule will be made more accurate and more efficient if their genetic origins are identical. If genetic identity cannot be achieved among propagule types, then at least near relatives should be employed, using sibs or offspring of the clonal propagules.

I've noticed a tendency at this meeting to call propagules of tissue-culture origin, of adventitious organ-culture origin, of normal-meristem organ-culture origin, of somatic-embryo origin, and maybe of other origins, "plantlets". If indeed there are consistent and significant differences among such propagule classes, this is going to confuse a lot of client users. And even if such differences don't occur, it would be useful to differentiate these while such a possibility is being tested. An attempt at such a nomenclature was made at the 1983 International Conifer Tissue Culture Work Group meeting in Tacoma, Washington, and was provisionally accepted by the Society of American Foresters Working Group on Tree Genetics and Improvement (1, 12).

4. PLANT MATERIAL USED DURING FUNDAMENTAL RESEARCH AND TECHNIQUE DEVELOPMENT

Here's where I'm an outsider. But it seems likely that researchers working at the fundamental-research and technique-development levels would find it useful to screen starting genotypes for clonal lines that handle well, and to then concentrate on just a few of the most promising of these. This is fine until they produce some independent "plantlets" from these lines. Then it becomes interesting and perhaps even important to find out if such "plantlets" grow like seedling-origin trees, or if not, how they differ from seedlings, and whether they might even be valuable candidates for deployment. Having produced an appropriate diversity of appropriate known origins makes life much better at this point.

5. RESPONSES TO A QUESTIONNAIRE

A questionnaire (Table 1) was distributed during the 1987 Grosshansdorf meeting, and all participants who are actively engaged in biotechnology, tissue culture and/or somatic-cell genetics completed and returned a copy. Of these, 19 are engaged in forestry projects, and one each works on coffee or fruit trees. The returns are summarized in part in Table 2.

5.1. Testing and deployment

Encouraging and perhaps surprising numbers of the participants reported that forest trees produced by tissue culture and/or by cell culture in their projects are being or have been field tested (42%). Three (16%) reported that such trees have already been deployed to production-forestry uses. Most (72%) expected such trees from their projects to be in the test and/or deployment phases within five

Table 1: Questionnaire

1. Has your project produced trees of tissue- or cell-culture origin that are being or have been field tested?
   ______ yes ______ no ______ If yes, please comment below

2. Has your project produced any trees (clones) of tissue-culture or cell-culture origin that have been deployed to production-forestry uses?
   ______ yes ______ no ______ If yes, please comment below

3. Do you anticipate, within the next five years, that some tissue-or-cell-culture-origin trees will be produced by your project that will be field tested or deployed?
   ______ yes ______ no ______ If yes, please comment below

4. Please indicate, with approximate percentages (10%, 25%, etc.), how much of the plant material currently being used in your project comes from the following sources:

   a. ___% Well-known clones, such as poplar clone I-214.

   b. ___% Well-tested clones that are currently being deployed to production forestry.

   c. ___% Clones that are currently under test but not yet deployed.

   d. ___% Well-known families, such as NZ 55x19 radiata pine.

   e. ___% Well-tested families that are currently being deployed.

   ___% full-sib families ___% female pedigree only (open-pollinated & polycross)

   f. ___% Families currently under test but not yet deployed.

   ___% full-sib families ___% female pedigree only (open-pollinated & polycross)

   g. ___% Families being neither field tested nor deployed

   ___% full-sib ___% female pedigree only (open-pollinated & polycross)

   h. ___% Material from a well-known provenance, such as the Darrington Douglas-fir provenance.

   i. ___% Material that is accurately source-identified, but whose performance is not well known.

   j. ___% Material known only to species, but without further identification.

5. Are you currently cooperating with ☐ physiologists, ☐ entomologists, ☐ pathologists, ☐silviculturists, ☐ nutritionists, or ☐ tree breeders, each of whom work at the whole-organism levels. Others ________________________________ .

6. How many genotypes are typically used at the beginning of one of your experiments? If different experiments use different numbers, indicate more than one option.

   ___ 1 ___ 2-10 ___ 10-100 ___ 100-1000 ___ >1000

7. Is your organization ☐Government ☐University ☐Private?

Table 2: A Partial Summary of Questionnaire Returns

Question (see Table 1)	Forestry % yes	Coffee & Fruit Trees % yes
1. Project's Trees Being or Have Been Field Tested	42%	50%
2. Project's Trees Already Deployed	16%	50%
3. Deployment Expected <5 years	72%	100%

4. Percentage of Project's Plant Material, by Category of Source (Forestry Only)

	% of Respondents Using Source	Of Respondents Using: Avg.% of Their Material	Of Respondents Using: Range of % Use
a. Well-known clones	21%	51%	25-100%
b. Tested deployed clones	21%	41%	10-75%
c. Clones under test, not deployed	32%	35%	5-100%
d. Well-known families	11%	12%	5-20%
e. Tested deployed families	26%	35%	5-67%
f. Familes under test, not deployed	32%	54%	5-95%
g. Families neither tested nor deployed	0%	-	-
h. Well-known provenance	32%	45%	5-100%
i. Accurately identified provenance	26%	30%	5-100%
j. Known only to species	32%	53%	2-100%

Percent of "4a", "4b" and "4c" clones from well-tested deployed families:
full-sib families: 40% open-pollinated & polycross families: 60%

Percent of "4a", "4b" and "4c" clones from families under test but not deployed:
full-sib families: 62% open-pollinated & polycross families: 38%

Percent of "e" families that are:
full-sib families: 80% open-pollinated & polycross families: 20%

Percent of "f" families that are:
full-sib families: 72% open-pollinated & polycross families: 28%

6. Number of Genotypes Typically Used When Beginning an Experiment

	1	2-10	10-100	100-1000	>1000
Forestry	4%	38%	42%	15%	0%
Coffee & Fruit Trees	0%	100%	0%	0%	0%

years. Such high percentages may merely indicate that the people and projects invited to this meeting are at present among the most successful. It surely indicates that attention to the "what next" questions associated with testing and deployment are timely. The higher percentages reported by the coffee and fruit-tree programs indicate their even more advanced states.

5.2. Genetic identification

With possible exceptions in "4c", "4f" and "4g", the higher one works in Table 2 "4a-h", the more efficient testing and subsequent deployment can be. It is heartening to see that many workers indicated substantial effort in the upper levels of these categories ("4a-e"). The coffee and fruit-tree projects reported 90 percent of their efforts in "4a". For those approximately 1/3 of the forestry workers who reported over half their material, on average, being in "4j", testing of any trees thus produced will be difficult, and confident deployment, if justified, could be much delayed. Class "4g", with no reported activity, is clearly underestimated. We (in Berkeley) have supplied both tissue-culture projects and biotechnological cell-culture projects with full-sib and open-pollinated families from genetic conservation collections, as an example. These are accurately source-identified and pedigreed, but are not under test.

For those using material with family pedigree available, it's interesting to note that those working at the clonal level use lower average percentages of the more efficient full-sib families (40% and 62%) than those working at the family level (80% and 72%). Those reporting family relationships at the clonal level were mostly working with *Populus*, *Juglans* and *Larix*, where polycrosses were apparently used to produce interspecies hybrids. There were also some interesting species-region differences within programs, such as 100% full-sib *Pseudotsuga menziesii* in the northwestern U.S., and 100% female pedigree (o-p and/or polycross) *Pinus taeda* in the southern U.S.

5.3. Genetic diversity

The number of genotypes produced for testing and ultimately for deployment is affected by and in some cases limited by the number of genotypes that enter an experiment. The design and nature of the tests responds to this factor (13). Furthermore, some laws or regulations governing deployment may require substantial minimum numbers of unrelated clones (15). The majority of questionnaire respondents indicated between 2 and 100 as the number of genotypes entering an experiment (Table 2). This level of genetic diversity is probably appropriate during the development of techniques, but it is likely to be too low as programs move toward deployment. The coffee and fruit-tree programs, working wholly at the 2-10 level of entry genotypes, are dealing with different expectations relative to management and risk, and should not be used as models for forestry purposes.

5.4. Whole-tree collaborators

Of the 21 respondents, three (all on forestry projects) indicated that they worked with no scientists at the whole-organism level. The other 18 listed colleagues in the following whole-organism fields, in order of frequency: tree breeders (13), physiologists (11), silviculturists (10), pathologists (8), entomologists (3), nutritionists and geneticists (2 each), soil scientists, practicing foresters and forest researchers (1 each). In addition, one each mentioned a molecular biologist, biochemist or biophysicist as being of particular importance as a collaborator.

5.5. Respondents' organizations

The 21 respondents were self-identified as 57% working for governments, 29% working at universities and 14% working in private industry or private practice.

6. A FEW SUGGESTIONS

6.1. Provenance

For species with few or no provenance tests, or whose provenance tests are recently established, it seems prudent to choose material for deployable manipulation conservatively. One would thus manipulate many trees from many local populations, anticipating eventual deployment of the genetically engineered products mostly or exclusively to the local regions from which they came. For that smaller number of species with many mature and well-executed provenance tests, trees from the broadly-adapted best-performing populations are probably the best candidates for biotechnological manipulation.

A common observation in provenance tests is that occasional individual trees from population-samples of below-average performance are among the very best trees in the test. These are sometimes chosen for further testing and for inclusion in long-term breeding programs. However, they are probably not good candidates for being gene recipients, nor should their biotechnologically modified products or immediate sexually propagated descendants be widely deployed to the region of the test. They may be good gene donors, at such time as we know enough to move desired genes from one tree to another.

6.2. Pedigree

Breeding programs often exist for our important species, and trees from their pedigreed families with one or more generations' performance records are particularly attractive as donors of tissue and offspring for biotechnological manipulation. Trees with full pedigree (the male and female parents are both known as to identity, relatedness, and native stands of origin) are preferred over trees with partial pedigree.

6.3. Clones

Clones that are well known (poplar clone I-214 is one such, although its parentage and origins are uncertain), clones that are being deployed based on completed or in-progress testing, clones that are being field tested, or clones that are being used in research by physiologists, silviculturists, etc., are all preferable as donors of cells, tissue or offspring, compared to clones without such a background of performance records.

6.4. The ideal

The best strategy when entering the field-testing and deployment phases is to draw material from many full-pedigreed clones that are already well tested, well known and widely deployed, and that are or have been used in various kinds of relevant research as well. This situation does not, to my knowledge, exist today. As clonal forestry develops, it will exist and should be exploited. Meanwhile, the more that is known about the material prior to its use in biotechnological manipulation, the better it is with respect to subsequent field testing and production-level deployment.

If unmanipulated material of the manipulated clones is not available for inclusion in testing, it may be useful for the biotechnology program to maintain and to supply unmanipulated close relatives of the engineered clones, for purposes of test comparisons.

6.5. Minority mixes

If one wishes to deploy promising novel clones before some combination of clonal history and current testing warrants confident commitment to widespread deployment, one may restrict the deployment of such material to some agreed-upon percentage of total planting. This might be done in intimate mixture, with the percentage of the new material pegged at a level so that the combined losses from the total failure of the novel clone(s) and the normal losses from the other trees would still result in an acceptable plantation. Or, if planted in a mosaic of monoclonal stands, the area committed to the new material might be pegged to acceptable losses of whole stands on an organizational or regional basis. This strategy seems the best available, but it is not fail-safe (13).

7. SOME WORRIES ABOUT PROFESSIONAL INFRASTRUCTURE

During the past half decade, as commitments to forest biotechnology research have been made, there has been in some countries (the U.S., at least) a substantial pulling back from more traditional forms of forest genetics and tree improvement. Whether the new is supported at the expense of the old or not is interesting but unimportant. What is important is that testing and tree-breeding are mixtures of science and informed art. With some exceptions, most good biotechnologists do not also have skills and training in field testing, nor the intimate experience with field-grown trees to spot unusual value or uncharacterized variation of a negative kind. Thus, when we need high-quality experienced field researchers to ask the "what next" questions and to take the next steps, they may be in short supply, or without organizational infrastructure to support the necessary field tests. This is an important concern if the sorts of work reported at this meeting are to contribute to our renewable forest resources, and not end merely in peer-reviewed publications in reputable scientific journals.

8. REFERENCES

1. BEY CF, DB HOUSTON, RJ DINUS 1986 Tree genetics and improvement. Part I. The new genetics. Jour For 84:34,36-37
2. Bolstad PV, WJ LIBBY 1982 Comparisons of radiata pine cuttings of hedge and tree-form origin after seven growing seasons. Silvae Genet 31:9-13
3. ELDRIDGE KG, DJ SPENCER 1986 Field performance of cuttings in Australia. Proc Workshop On Growing Radiata Pine From Cuttings. NZ For Res Inst Rotorua. pp 9-10
4. FRAMPTON LJ Jr 1986 Field performance of loblolly pine tissue culture plantlets. IUFRO Conf on Breeding Theory, Progeny Testing and Seed Orchards, Williamsburg VA. pp 547-553
5. HUEHN M 1985 Theoretical studies on the necessary number of components in mixtures. 1. Number of components and yield stability. Theor Appl Genet 70:383-389
6. HUEHN M 1986a Theoretical studies on the necessary number of components in mixtures. 2. Number of components and yield ability. Theor Appl Genet 71:622-630
7. HUEHN M 1986b Theoretical studies on the necessary number of components in mixtures. 3. Number of components and risk considerations. Theor Appl Genet 72:211-218

8. HUEHN M 1986c Theoretical studies on the necessary number of components in mixtures. 4. Number of components and juvenile-mature correlations. Theor Appl Genet 73:53-60
9. HUEHN M 1987 Clonal mixtures, juvenile-mature correlations and necessary number of clones. Silvae Genet 36:83-92
10. KRUSCHE D 1982 Host-parasite interaction and the number of clones in a clonal mixture. In J Kleinschmit, ed, Proc IUFRO Joint Meeting Working Parties on Genetics about Breeding Strategies including Multiclonal Varieties, Sensenstein. p 171
11. LIBBY WJ 1982 What is a safe number of clones per plantation? In HM Heybroek, BR Stephan, K von Weissenberg, eds, Resistance to Diseases and Pests. Purdoc Wageningen, The Netherlands, pp 324-360
12. LIBBY WJ 1986 Clonal propagation. Jour For 84:37-38, 42
13. LIBBY WJ 1987 Testing and deployment of genetically engineered trees. In JM Bonga, DJ Durzan, eds, Cell and Tissue Culture in Forestry. Martinus Nijhoff, Dordrecht, pp vii, 167-197
14. LIBBY WJ, RM RAUTER 1984 Advantages of clonal forestry. For Chron 60:145-149
15. MUHS HJ 1987 Rules governing the release of forest reproductive material derived by in vitro culture and breeders' rights. This Proceedings
16. RAUTER RM 1982 Recent advances in vegetative propagation including biological and economic considerations and future potential. In J. Kleinschmit, ed, Proc IUFRO Joint Meeting Working Parties on Genetics about Breeding Strategies including Multiclonal Varieties. Sensenstein, pp 1-26
17. ROULUND H 1981 Problems of clonal forestry in spruce and their influence on breeding strategy. For Abstr 42:457-471
18. SMITH DR 1983 Field evaluation of radiata pine micropropagules. In Abstracts of The International Conifer Tissue Culture Work Group, Weyerhaeuser Co, Tacoma Wa USA, p 29

RULES GOVERNING THE RELEASE OF FOREST REPRODUCTIVE MATERIAL DERIVED BY IN VITRO-CULTURE

H.J. MUHS

Federal Research Centre for Forestry and Forest Products, Institute of Forest Genetics and Forest Tree Breeding, Sieker Landstr. 2, D-2070 Grosshansdorf 2, Federal Republic of Germany.

0. ABSTRACT

The international rules, mainly the OECD-Scheme and the EEC-Directives do not have special rules for the marketing of forest reproductive material derived by *in vitro*-culture, but both the OECD and the Commission of the EEC are planning to amend their rules. The principle of approval is described, which points out that the basic material requires approval but the reproductive material derived from the basic material moves in the market. The problems for approving and marketing *in vitro*-culture derived plants are discussed and a solution to the problem is proposed. In case of clonal material, which can be produced either by macro- or micropropagation, two countries, Sweden and the Federal Republic of Germany, have set up rules for marketing multiclonal mixtures. Each country has followed different sets of approvals for the marketing of multiclonal mixtures. These are discussed in a brief form, focusing on the main features of the rules.

1. INTRODUCTION

Since the seed trade had expanded and the knowledge about genetic variation and adaptation of forest tree species had increased, it became necessary, to formulate rules which should protect the forests regenerated from trees raised from unsuited seed sources. This process started in the 1920ies of this century in Germany and some other Central European Countries. In 1957 the Federal Republic amended the rules, in which the minimum requirements for the approval of basic material, the sealing and labelling of reproductive material and other rules were laid down. While in the early (before 1957) regulations, the employment of the reproductive material was ruled, the last directive sets up rules for the marketing of reproductive material. This principle has been adopted by the international rules, the EEC-Directives (1,2,3) and the OECD-Scheme (4). Both are aimed at an international market, which shall be stimulated by harmonizing the different national rules in case of the EEC-member countries or by applying the OECD-Scheme, if material moves in international trade between OECD-member countries, which are not EEC-member countries simultaneously. While the EEC-Directives are based on two categories of reproductive mate-

M. R. Ahuja (ed.), Somatic Cell Genetics of Woody Plants, 211–219.

rial, "selected" followed by "tested reproductive material", the OECD-Scheme distinguishs between 4 categories, mainly the "source identified reproductive material", "selected reproductive material","reproductive material from untested seed orchards", and the "tested reproductive material". Generally the categories "selected" and "tested" in both rules are identical, while the categories "source identified reproductive material" and that from "untested seed orchards" are missing in the EEC-Directives, that is, reproductive material of these categories cannot be marketed. A brief description of both rules are given elsewhere (6,7). In this paper a special question shall be analysed, which is the following: Are there rules governing the release of forest reproductive material derived by _in vitro_ -culture on an international or national level and how should they be fixed? The first question is easily answered: There are no special regulations either in the EEC-Directives or in the OECD-Scheme, which are considering reproductive material derived by _in vitro_-culture. Up to now only two countries, the Federal Republic of Germany and Sweden, have set up rules for the marketing of clones and clonal mixtures for species other than poplars (8). But no special reference is made to material derived by _in vitro_-culture. The second question is rather difficult to answer. Some reflections are given below.

2. THE PRINCIPLE OF APPROVAL

Before going into detail it seems to be necessary to reflect on the principle of approval , which is the basis of marketing reproductive material and which is common to all rules referred to in this paper. Figure 1 shows the main components of the principle of approval. The basic material, which can be a seed stand, a seed orchard, a clone or a clonal mixture, is the object of an official act called approval.

The approval gives permission to the owner of the basic material to collect reproductive material from the basic material and sell it in the market. It is a common understanding that from seed stands and seed orchards only seed or wildgrown seedlings from that stand or orchard should be collected as reproductive material coming from that stand or orchard. It is not allowed to market other reproductive materials from a seed stand or a seed orchard like vegetative material to produce rooted cuttings or other. The case with clones and clonal mixtures is analogous. It is permitted to market vegetatively propagated material or plants derived from an approved clone or clonal mixtures but not seed or seedlings originating from those clones or clonal mixtures.

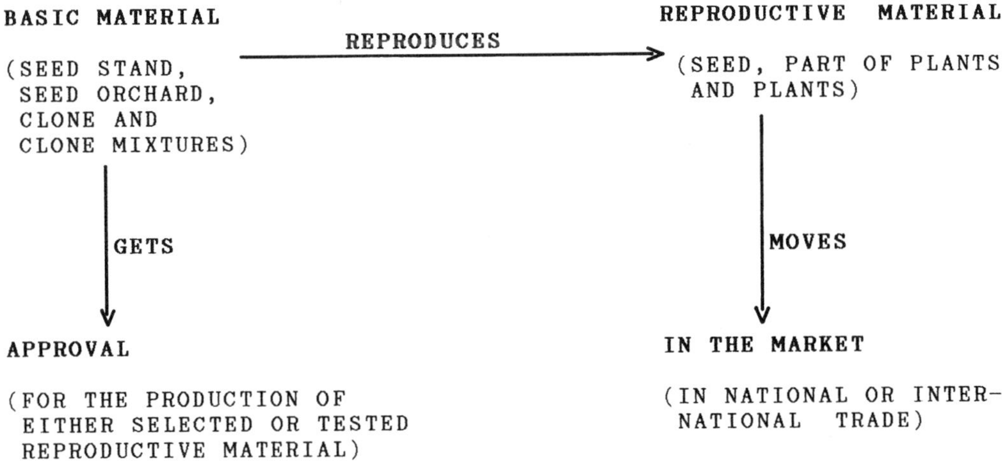

FIGURE 1. The principle of approval

Thus the principle of approval is based on:

- the physical existence of the basic material in a seed stand of a given size or a seed orchard of a given composition or a specimen of a clone in a clone bank or a given set of specimens of clones for producing clonal mixtures.
- the mode of reproduction, either sexual or asexual (vegetative)

Approval can be granted for the production of either selected or tested reproductive material. This is the case with the EEC-Directives and the German national law for the marketing of forest reproductive material (5). While the basic material must get approval, the reproductive material does not require approval. Thus all reproductive material that moves in national or international trade is not approved, but their basic material from which it originates is approved. Consequently the reproductive material should not serve for the production of reproductive material itself except when it has been approved as basic material. The terms reproductive material and reproduction are used for:

- sexually derived plants in case of seed stands and seed orchards (sexual propagation)
- asexually derived plants in case of clones or clonal mixtures (vegetative propagation).

The terms do not distinguish between the reproduction and the multiplication of the reproduced material in a strict sense. Thereafter reproduction of parent trees means that the seed derived have undergone meiosis followed by fertilization, while reproduction of an ortet results in ramets, which have been produced via mitotic cell division only without undergoing meiosis. The term bulk propagation or bulking of basic material usually combines both ways of reproduction. The basic material, for example a seed orchard, produces seed, which is multiplied by vegetative propagation of cuttings taken from the young seedlings derived from that seed. The main feature of bulking is that the identity of the plants multiplied gets lost. So far bulking is not allowed according to most rules.

3. THE PROBLEM OF APPROVING *IN VITRO*-CULTURED PLANT MATERIAL AND THE SOLUTION PROPOSED

The problem is that the mode of reproduction varies between the *in vitro*-cultured plant material using different techniques. Some examples of *in vitro*-cultured plant material derived by different modes of reproduction are given below:

- embryo culture; if the basic material is known and the *in vitro*-technique aims at raising only one plant from a seed, the mode of reproduction can be compared to that of conventionally derived and raised seedlings, but if the aim of embryo culture is to multiply the embryos, the mode of reproduction is comparable to bulk propagation

- clonal micropropagation by using organs, tissues, cells or protoplasts; this mode of reproduction is the same as vegetative propagation

- selection and propagation of variants from cultures showing somaclonal variation; in this case the source of variation is not known, which may be different kinds of mutation. The variants are genetically different from the basic material and multiplied vegetatively thereafter

- haploid culture; plant material coming from haploid cultures are genetically different from the diploid basic material, the material has undergone meiosis but this is not followed by fertilization. The occurence of chimeras is possible and the propagation is done by vegetative multiplication

- cultures of diploids and polyploids by doubling or fusion. In this case the plants derived are genetically different from the basic material. During mass propagation by vegetative multiplication changes in the chromosomal status can occur

- cybrids or plantlets derived from cells with additional organelles transferred are genetically different from their basic material in their composition of extranuclear genes; mass propagation has to be done by vegetative multiplication

- genetically engineered material is different from its basic material and needs to be vegetatively multiplied for propagation.

It is shown that *in vitro*-culture derived plants may be obtained by different modes of reproduction, which results in different reproductive material. This is advantageous in breeding but disadvantageous for getting approval. At the present time the marketing of the *in vitro*-culture derived plant material, is not allowed. But the OECD-Scheme and the EEC-Directives should include a regulation for marketing plants derived from *in vitro*-culture. Discussions have already been started at different levels.

An approach to a solution of the problem may be the following. The process from the basic material through *in vitro*-culture methods to get propagules should be separated into a breeding phase and a mass propagation phase. The first one should not be included in the rules govering the release of reproductive material. Subsequently the term "basic material" should be split into

- basic material (breeding term = b.t.)
- basic material (legal term = l.t.)

Then the process described above looks as follows:

- during the breeding phase the basic material (b.t.) will be treated by the *in vitro*-culture methods resulting in a product which can be regenerated to a plant. After having tested the regenerated plants, they can serve as basic material (l.t.).

- during the mass propagation phase plants are produced by vegetative propagation using the basic material (l.t.). The result is a clone.

The consequence of the proposed solution is the exclusion of the breeding phase from being regulated by the rules for marketing forest reproductive material. Except in case of the embryo culture all *in vitro*-culture derived plants can be handled as clones, because they are mass propagated by vegetative means. In this case the basic material (l.t.) must fulfill the requirements for getting the approval as outlined above. Further they must prove their stability and trueness-to-type. They have to be tested in comparison to one or several standards before getting the approval, the conditions of the test are laid down in the rules. They have to be marketed as clonal mixtures according to the German rule if species other than poplars are in question. If this is the rule, they have to fulfill several additional requirements, which are described below.

4. RULES FOR THE MARKETING OF MULTICLONAL MIXTURES

So far only two countries have set up rules for multiclonal mixtures in their national law for marketing of forest reproductive material, namely Sweden since 1982 and the Federal Republic of Germany since 1979. In the following synopsis (Table 1) the rules and some major requirements are compared, which in this paper can be described only in a very preliminary and incomplete way. But it shows the different regulations of approval for the same problem.

TABLE 1. A synopsis of Swedish and German Laws governing the marketing of forest reproductive material

Subject	Sweden	Fed. Rep. of Germany
1) Legal status of the rules	law since 1982	law since 1979, details are laid down in a General Administrative Regulation, amended in 1985
2) Selection of basic material	not regulated generally, bulks derived from approved basic material are allowed	not regulated
3) Multiclonal mixtures, mosaics, or single clones for species other than poplars	multiclonal mixtures	multiclonal mixtures only
4) Species covered other than poplars	Conifers	all species (18 totally under the law, and 19 if the genus _Populus_ is included)
5) number of clones per mixture	according to test levels, level 1; 120 clones level 2; 60 clones level 3; 30 clones	dependent on species; 500 and 100 respectively for species used for large area reforestation. 100 and 20 resp. for species used for common sites. First number is the minimum number for common sites, second number is the minimum for special sites.
6) description of the identity of clones	the producer is obliged to keep a record on origin or descent for single clones and for bulks	description of each clone according to provenance or origin or, if it is a breeding product, descent and some additional items

TABLE 1. (continued)

Subject	Sweden	Fed. Rep. of Germany
7) use of untested clones	untested clones may be used for examination generally up to 1989	not allowed
8) testing procedure: - single clone or mixture testing	single clone	single clone
- duration of test	according to test level level 1: 6 years level 2: 9 years level 3: 6+12 years (a replication in time of test at level 1)	dependent on species from about 10-30 years commonly used in progeny testing, (there are no minimum years of testing given)
- number of test sites	level 1: 2 sites level 2: 4 sites level 3: 6 sites	at least 3 sites
- traits to be analysed	climatic suitability and growth rate	identification, adaptation, growth vigor, resistance
- control (standard)	controls for test at level 3 shall be selected in consultation with the National Board of Forestry	detail specification for each species in a list of "Standards"
- evaluation of results	general results must be published	procedure has to follow special rules (the superiority must by significant at the 95 % level)
- who is the tester	producer of the mixture	producer
9) bulk propagation	- not more than 100 ramets from each plant derived from seed may be bulked up	not allowed

TABLE 1. (continued)

Subject	Sweden	Fed. Rep. of Germany
	- bulk-propagation of mixed clone-identified material is only permitted for clonal mixtures approved at the level 2 and 3 and in only one propagation cycle	not allowed
10) mixing procedure, clonal identity	clone identity is maintained in all the following propagation cycles and lost only when material is combined in mixtures for use in forestry	the same as for Sweden
11) restrictions for use	mixing before use, applications shall specify the seed orchard zone in which the material shall be used	mixing before use, if tested for special sites, this shall be specified on the label
12) approval	by the National Board of Forestry	by the authorities of the Federal Lands as "tested reproductive material"
13) limitation of approval	according to test level, level 1: 250.000 level 2: 750.000 level 3: 1.500.000 cuttings per clone. These amounts shall include cuttings used at lower test levels	10 years, prolongations for further 10 years possible
14) registration of clones	to be done by the producer	to be done by the resp. authority
15) control	to be done by the National Board of Forestry	to be done by the resp. authority

There are some more informations about clonal mixtures (8,9), which may be of interest in this context. If the rules for marketing clonal mixtures can be harmonized in that way, that the OECD-Scheme and the EEC-Directiv will be basicly identical and will include all the aspects discussed above, the international market will gain a lot.

5. OUTLOOK

After the in vitro-culture derived plants have proven their suitability in forestry, they should have the chance to move in the market. In this case international and national rules should be set up to allow this. At the moment a discussion on different aspects of this problem has been initiated in the IUFRO Working Party on Legislation of Forest Reproductive Material. The OECD as well as the EEC are going to amend the scheme, the directive, respectively, which may take some years at least. Beside this it must be decided at the Government or even Parliament level, whether to allow the release of genetically engineered plantlets and plant them into the field. By now this is not the case.

REFERENCES

1. EEC (European Economic Community): Council Directive of 14 June 1966 on the marketing of forest reproductive material (66/404/EEC). Official Journal of the European Communities. 2326/66-11.7.66, p.161-167,1966.
2. EEC: Council Directive of 18 Febr. 1969 amending the Council Directive of 14 June 1966 on the marketing of forest reproductive material (69/64/EEC). Off. Journ. of EC no. L 48/12-26.2.69, p. 61, 1969.
3. ECC: Second Council Directive of 26 June 1975 amending Directive No. 66/404/EEC on the marketing of forest reproductive material (75/445/EEC), of L196/14, p. 14-23, 1975.
4. OECD (Organisation for Economic Cooperation and Development, Directorate for Agriculture and Food): OECD Scheme for the Control of Forest Reproductive Material Moving in International Trade, C(74)29 Final. 5th March, 1974, Paris, pp. 24,1974
5. FSaatG: Gesetz über forstliches Saat-und Pflanzgut vom 26.7.1979, Bundesgesetzblatt, Jahrg. 1979, Teil I, S. 1242-1261 (eingeschlossen die Anlagen I bis VII).
6. Muhs, Hans J.: Identifying problems and legal aspects of using multiclonal mixtures. In Proc. of the JUFRO Joint Meeting on Breeding Strategies including Multiclonal Varieties. Publ. by Lower Saxony Forest Research Institute D-3113 Staufenberg-Escherode, FRG, p. 92-108, 1983.
7. Muhs, Hans J.: Legal aspects concerning the forest reproductive material. In: Proceed. of the 18th JUFRO World Congress, Ljubljana, Sept. 86, Div. 2, Vol. I, 100-111, 1986.
8. Muhs, Hans J.: A synopsis of some national regulations for multiclonal mixtures as applied for Norway spruce and species other than poplars. Mitt. d. Bundesforschungsanstalt für Forst- und Holzwirtschaft, Nr.154, S. 202-209, 1986
9. Muhs, Hans J.und Krusche, D.: Klonmischungen. AFZ, Dez. 1986, Heft 51/52, S. 1316-1319, 1986.

GUIDELINES FOR RESEARCH AND DEVELOPMENT IN BIOTECHNOLOGY

STANLEY L. KRUGMAN
TIMBER MANAGEMENT RESEARCH
U.S. FOREST SERVICE
WASHINGTON, D.C. 20090-6090 USA

ABSTRACT

The new biotechnology offers an array of technics for overcoming the conventional biological constraints of the large size of trees and the delayed sexual processes common to woody species. By providing mechanisms for identifying, isolating and transferring single or complex genetic units it is now possible to accelerate tree improvement activities. However, this new ability to rapidly move genes between and among like and unlike organisms has created concern and debate among scientists and the public. To address these issues the U.S. Department of Agriculture has developed a set of research guidelines for the management of the new biotechnologies. The basic concept of the guidelines is that the end product of genetic engineering is no different from those developed by traditional genetics and that there are safe and responsible means of conducting biotechnical research. In the fall of 1987 detailed research guidelines for laboratory research, containment and field testing genetically engineered organisms, including trees, were made available. If followed the research and testing of genetically engineered trees will pose no hazard to the public or the environment.

1. INTRODUCTION

The new biotechnology is unlike any biological science methodology that we have ever had to cope with. The new techniques of biotechnology offers forestry a set of working tools for identifying, isolating and transferring single or complex genetic units from one organism to a second organism. The organisms need not to be closely related in order to exchange genes. For forestry this tool offers a unique opportunity for overcoming our conventional biological constraints of large size and delayed sexual processes. There are those who feel that these new techniques offer the greatest single-known biological technology for increasing growth productivity by its ability to introduce single, useful genetic traits without disrupting desirable existing features. At this stage of development forest biotechnology is a high risk investment, but one that is balanced by a very high payoff potential. In forestry, we have already seen the new biotechnology focus on such important areas as: gene location, herbicide and disease resistance, physiological stress and photosynthesis.

But the new science has also placed us in the forefront of a new arena; i.e. ethics, public relations and environmental concerns. This new technology has generated a great deal of concern and debate in both science and in

M. R. Ahuja (ed.), Somatic Cell Genetics of Woody Plants, 221–225.

society. A number of serious ethical questions have been raised and must be addressed. These issues have been under discussion in the United States for the last 12 years. We have attempted to construct research guidelines and regulations that are good and responsible science, and that are responsive to the needs of society at large. The main aim of these research guidelines are to avoid potential hazards to society and of course, other biological systems. These are the main concerns of society.

The new science of biological engineering, or biological tinkering, depending on your viewpoint, has brought out some of the best and worse thinking in terms of science development, and science law. In Europe there are new laws already passed or being planned that would bring the development of the new biotechnology to a halt. This is unfortunate since with any new technology there are various levels of complexities and hazards which must be recognized and separately managed. There remains a great deal of misinformation concerning these new technologies. For this, I must blame the scientific community and the popular press. If there was ever a case for open communication among scientists and the general public, the story of the new biotechnology is a perfect example. In the United States, the new biotechnology arose during the 1960's and early 1970's from the unique merger of traditional and molecular genetics with biochemistry. It came during a period when the traditional ways of doing science were being seriously challenged, in part, because of environmental concerns related to pesticides and because of the role that science was playing in an unpopular war in Asia. Collectively, these events clouded the understanding of the potential as well as the limits of science. Some felt that with this new technology no one or thing was safe from scientific tinkering since the traditional barriers of gene transfer were now broken. Time does not permit a detailed discussion, but needless to say the idea of releasing genetically engineered organisms to a natural environment was frightening to many then and still today. And there are, of course, legitimate concerns that must be considered when known pathogens are involved in the research.

We began in the mid-1970's to develop a responsible set of research guidelines for the plant and animal agricultural sciences to address the concerns voiced by the public and scientific communities. At the same time, the medical sciences had already begin the development of guidelines for more basic research in genetic engineering. In the last few years, we have enlarged our basic guidelines to cover the areas of containment and release of genetically engineered agricultural plants and animals. No attempt in this brief report will be made to cover the technical details since during the fall of 1987, the complete research biotechnology guidelines for the United States will be published. It will be a document of several hundred pages covering all forms of biological organisms from higher plants and animals to microorganisms. It will provide an array of tested laboratory and field procedures for genetically engineered organisms. Included in the guidelines will also be recommendations for several non-genetically engineered technologies such as cell fusion and somaclonal techniques. This we felt was necessary because the public placed these methods with genetic engineering. It is my hope that your awareness of these guidelines will be of assistance to you as you address these same issues in your own countries. It has been our experience that the scientific community must play an active role with the politicians and concerned citizens in the managing of modern science. There is no other choice if you expect new scientific fields to be developed and employed to meet the needs of current and future societies.

2. PHILOSOPHY OF THE RESEARCH GUIDELINES

The basic policy of these guidelines rests on the assumption that the safe conduct of forestry research involving biotechnology resides as a matter of obligation principally with the individual who conducts the research activities. Expertise, motivation and good judgment are the key essentials to protection of health and the environment.

A second major concept of the guidelines is that the end product of genetic engineering essentially is no different from those developed by traditional genetics. A genetically modified and improved _Populus_ seedling is no different from those improved by traditional selection and breeding. However, the process that leads to the engineered tree is what we propose to regulate, i.e. research guidelines. These guidelines are subjected to continued review, updating and revision. It is the responsibility of research institutions and scientists conducting biotechnological research to adhere to the intent of the guidelines, as well as to their specific provisions, but also to contribute to their continued review and improvement. As noted earlier, the scope of the guidelines covers those processes which cause heritable variation in biological organisms through the process of causing heritable variation through the manipulation of recombinant DNA, recombinant RNA, specific molecular gene vectors, cell fusion, or other non-classical genetic procedures conducted at the cellular or sub-cellular level.

3. REVIEW PROCESS

Because of the dynamic nature of the research guidelines, a review and approval process has been established. At the local laboratory level, Institutional Biotechnology Committees (IBC), have been established to review all biotechnological investigations and approve certain none to low risk experiments requiring limited release into the environment. These are defined in some detail in the guidelines and have been extensively evaluated over the 12 years as to their potential hazard to the environment.

For potentially moderate risk requiring limited release into the environment experiments, and these too are defined in detail, they must be approved by a national Recombinant DNA Advisory Committee (RAC) established by the U.S. National Institute of Health or by the U.S. Department of Agriculture. Recently a series of review committees coordinated by the Office of Agricultural Biotechnology (OAB), have been established in the U.S. Department of Agriculture to review these types of experiments and those involving potential release in which research experience is limited or might pose an unknown hazard to the environment. These review committees are composed of members from all sectors of the biotechnology and science communities including federal, private, university and public institutions.

There are, of course, a number of prohibited experiments. These deal with known and proven severe hazards to human and plant health. These have been clearly identified and outlined in some detail. This list does change as we learn more about the technology and the organisms we work with. However, experience over time has demonstrated the wisdom of not initiating a whole range of investigations, even under the highest level of containment. For example in terms of release to the environment all engineered microorganisms involving genetic inserts or constructs toxic to vertebrates are prohibited.

4. CONTAINMENT

A major concern has been the containment of genetically engineered organisms. A great deal of effort has gone into studying various levels of containment for all types of experiments. The results of these studies have now been placed in the new guidelines. Obviously, physical containment conditions within a given laboratory is not always appropriate for all organisms. For genetically engineered higher organisms such as trees, information on containment does exist and has been tested. For plants, including trees, we have identified three biosafety levels.

Biosafety Level 1 is designed to provide a minimal level of containment for specific recombinant DNA research. This would include research involving trees on other plants for which there is no recognizable and predictable risk to the environment in the event of accidental release or for which there is convincing biological evidence that precludes the possibility of survival, transfer or dissemination of the recombinant DNA molecules into the environment. This level represents ordinary greenhouse or growth chamber facilities, but does incorporate accepted procedures for good pest control and growing conditions. Such facilities will provide a degree of containment that adequately controls sexual and vegetative production, and minimizes the potential for release of biologically viable plants, plant parts and microorganisms associated with them.

Biosafety Level 2 is established to provide greater containment of plants and certain associated organisms and is recommended for those experiments in which there is a recognized possibility of survival, transmission or dissemination of the recombinant - DNA - containing organism but the consequences of such a release has a predictably minimal biological impact in the event of inadvertent release. As with Biosafety Level 1, Biosafety Level 2 relies upon accepted scientific practices for conducting research in greenhouse's with organisms infecting or infesting plants in a manner that minimizes or prevents inadvertent contamination of other plants within or surrounding the greenhouse.

Biosafety Level 3 is the highest level of containment for research with plants associated with certain pathogenic organisms. Biosafety level 3 containment prevents and eliminates cross contamination within the containment system.

5. RELEASE

The new biotechnology was initially developed in the medical sciences. As such, regulation and guidelines were constructed to consider the laboratory processes involved since the intent of the new science was essentially in the development of new and improved products from genetically engineered organisms, not the organisms themself. In forestry, we are interested in products from genetically engineered organisms; but, in addition, we are also interested in developing superior genetically improved trees. To use these new whole tree products they must be tested and grown in an array of external natural environments to determine their value. Thus it has become necessary to develop additional guidelines for an orderly and safe release of genetically engineered plants. Discussions of the last few years related to release have dealt with microorganisms, rather than higher plants. This is due in part because more is known about higher plants in their various natural or man-made environments, than about microorganisms. Higher plants; i.e., trees, are easier to control in the environment.

Basically, the principles for containment involving release of a genetically engineered tree is directed toward preventing release, disposal and establishment of plants by means of seeds, vegetative propagules and pollen during the testing period. Since plants vary, the actual field containment depends on the biology of the plant and the nature of the experiment.

A set of general guidelines have been developed so that plants can be field tested without undue risk to the environment. Two types of field containment is recognized; i.e., physical and biological.

Physical containment can include physical structures or other practices including geographical or climate. Physical containment can also include various techniques such as soil sterilization or herbicide application.

Biological containment of plants can include genetic modification of the plants which reduces the ability of the tree to survive in the environment or reduces its ability to survive or reproduce. For example, isolation can be achieved by removing all cross-compatible plants within the known effective pollinating distance of the particular species. Pollen release can be prevented by blocking pollen dispersal. The same can be done for seeds. Once again, the point of this activity is to prevent the escape and reproduction of the genetically engineered tree until it is adequately field tested to ensure it is not a hazard to the environment. For trees, I do not see these procedures as a major constraint for adequate field testing.

At the conclusion of any experiment all field material must be destroyed unless it is demonstrated as being safe. I would expect our experiments to be somewhat longer than similar experiments by our colleagues in Agriculture. In order to test a forest tree, we must both test the value of the transformed gene as well as the adaptability of the new product.

6. CONCLUSION

For many this brief discussion of guidelines appears to be overly complicated for the new forestry biotechnology. The simple single gene systems we are attempting to bioengineer do not appear to be any different from the gene systems that we employ in traditional breeding. Yet as I noted earlier, we need to address the concerns of the general public. We must demonstrate that our science is in fact safe. For that reason, we will continue to over design our experiments until we have a clear safety record in forest biotechnology. During 1988, the U.S. Forest Service will release the first genetically engineered tree for field testing. We have now began the paper process, which will include an environment review and assessment. In developing our first field test we will need to inform the public, but also those in forestry who are also unsure of this new technology. Needless to say, in the next few years our biotechnology research guidelines will be sorely tested. I am optimistic that both the scientific and public communities will accept our guidelines as being responsible and we can get on with the research and development of improved forest trees.